ENVIRONMENT AND DEVELOPMENT CONUNDRUM

Environment and Development Conundrum

Edited by
Kumar B. Das

1992
Discovery Publishing House
NEW DELHI-110002

First Edition 1992

ISBN 81-7141-178-9

Published by : DISCOVERY PUBLISHING HOUSE
4594/9, Darya Ganj, New Delhi-110002

Printed in India at Sangita Printers, Maujpur Delhi-110053

NIRA AND ABHAYA
the real force behind

Preface

The interaction of ecology and economy are becoming very clear everyday. Ecologically overstressed societies are further impelled to more intensification of pressure on their environment in an attempt to survive. Now we have a rude awakening and a jolting realisation of what have been done to the environment and how man tends to jeopardise its future. The consequence of the vertical economic growth, is obviously an ecological disaster. Therefore, we urgently need to develop an ecosystem approach to the management of resources. This approach is not anti-life. Its fundamental objective is total well being of the society for the present generation and for the future generation. This can only ensure a sound and sustainable balance between a productive environment and a protective environment. Economic development can not subsist upon a deteriorating environmental resource base. No country or no economic agent should be allowed to exploit its comparative advantage in pollution—intensive or energy-intensive activities, Discrete approach is no longer appropriate and must be replaced by a holistic approach that considers man as a part of but not apart from the environment. We have to place the preservation of fragile eco-system ahead of technology, population and economic development. Notwithstanding the practical difficulties we should be concerned about the problem in its proper dimension and severity. In view of protection and preservation, the negative effect of the rise in material welfare requires reorientation with respect to certain prevailing economic tendencies so that in addition to economic development, the protection of environment and preservation of natural resources have to be conceived of as an explicit aim in the planning and political decision making.

This book embodies research papers, contributed by reputed research scholars from different institutions. They make diagnostic probings and offer their practical insights into the causes and consequences ecological crisis and environmental deterioration. Tweleve eminent experts offer a set of suggestive measures for improving the environmental quality and resources base for the posterity. They emphasis the need to have an ecosystem approach as an alternative, which is necessarily an intergenerational approach, in order to have a sustainable growth process.

This book could be published only due to the cooperation and interest of some of the scholars experts really concerned about the problem of ecological crisis of the world. I am very much grateful to all the paper contributors for their encouragement and ready response far bringing out this volume. I am extremely indebted to Professor Dr. Brian Clark of University of Aberdeen, U.K., and Professor Larry Canter, of University of Oklahama, USA for their kind guidance and inspiration for bringing out this volume. I feel equally indebted to Professor Peter Nijkamp of Vrije University of Amsterdam and Professor Peter Frederick of Bamberg University West Germany for their perennial patronage and motivation for me to specialise in this branch of knowledge. My thanks are also due to Professor G.S. Bhalla, Jawaharlal Nehru University, New Delhi, Professor C.R. Pathak, I.I.T., Kharagpur, Professor C.S. Mishra, Ravisankar University, Professor B.C., Mehatta, Sukhadia University, Rajsthan, for their continuous encouragement.

Lastly I express my thanks to Shri D. Mahakul for his excellent typing and to Shri T.R. Wasan for bringing out this book in a record time with a beautiful get up.

Kumar B. Das

Contents

1

Introduction

The word 'ecosystem' comes from two words, "Ecology" and "Systems". Ecology comes from the greek word, "o i k o s" meaning households is considered the study of the "economy of nature". The word "Ecosystem" was coined by A.G. Tansley back in 1935. Ecology is the biology of ecosystems. But ecosystem refers to the interaction, inter-dependence, an orderly working totality of complex factors constituting the total environment.

The natural ecosystem consists of the physical environment, inhabitated by plants, animals, each of which in turn modifies the climatic, the hydrological and the nutritional aspect of the environment. Each group of species is made up of individuals, collectively a population, that are held together by some form of social and biological interaction. These interacting man, animals, plants and the environment make up the ecosystem. It functions by maintaining a flow of energy and a cycling of materials, within a structural frame-work of the environment and vegetation. It is an unit of biological organisation made up of all of the organisms in a specified space interacting with the physical environment so that a flow of energy leads to characteristics tropic structure and material cycles within the system. Inherent in the concept of ecosystem is the idea of stability and homeostatis. Stability implies persistence; the system remains much the way it was. In natural ecosystem this stability means that the system can survive many changes, but still preserve some recognizably similar structure. To effect this stability some homeostatic or feed back is necessary.

Processes in the Ecosystem

The rapidly growing urban decay does not just happen. Our environmental problems and our population problem have not appeared overnight. They have been a long time in the making. Ecosystem functions as a complex of interacting systems involving the living elements and the non-living. Due to early emergence and evolution of man he gradually evolves methods of channellising energy of natural ecosystem into artificial ecosystem to support his own population and animals. His ability to channel energy for his own use changed man's attitude towards the land and environment, fostered the development of urban life and encouraged technology through which man caused disequilibrium in the ecosystem. The ability to increase the carrying capacity of the ecosystem permitted an expansion of the human population which has reached a level at which it is placing too heavy a demand on the complex systems of the ecosystems and is endangering the human habitat.

The urban ecosystem takes the form it presently has, has had in the past, and will have in the future, because of the operation of certain ecosystem processes that permeate its growth. There are mainly eight types of ecological processes. Those are: (1) concentration, (2) centralisation, (3) decentralisation, (4) segregation, (5) specialisation, (6) invasion, (7) routinization and (8) succession. These processes need not be discussed seriatim. These are in fact operating in an urbanisation process concurrently perhaps at different rates of speed under different circumstances with one being much more pronounced in its effect than another. Whatever may be the ecological process of work, there are certain background factors like competition and mobility which affect their operation. The competition factor is apparent in all of the processes. Competition for space derives from the fact that two objects cannot occupy the same space at the same time. It may be said that to a large extent the structure of an urban area—the distribution of its population, its land use pattern—is a product of the competitive interaction of different population groups, landuses, functions and institutions. Mobility is a measurable movement in the temporal or spatial framework, specifically the movement of people and goods, land uses and functional areas from one part of a city to another. It is so common to most of the processes that virtually none can be comprehensible unless mobility is considered as a component.

Ecological Succession

The climax of the invasion process, where the original population group or use pattern is gradually replaced is termed as the ecological succession of the ecosystem. It is defined in terms of 3 parameters. (1) it is an orderly process of community development that is reasonably directional and therefore predictable, (2) It is controlled by the community even though the physical environment determines the pattern, the rate of change and often sets the limit as to how far the development can go, (3) It culminates in a stabilized system in which maximum biomass and symbiotic function between organism are maintained per unit of available energy flow.

The trends expected in the process of ecological succession can be found out by making a contrast between immatured and matured ecosystems.

Ecosystem Attributes	Young Stage	Mature Stage
1. P—R ratio	> 1 <	= 1
2. P—B ratio	High	Low
3. B—E ratio	Low	High
4. NCY	High	Low
5. Food Chain	Linear Predominantly grazing	Weblike Predominantly detritus
6. Total Organic material	Small	Large
7. Niche specification	Broad	Narrow
8. Production	Quantity	Quality
9. Internal symbiosis	Under-developed	Developed
10. Nutrient conservation	Poor	Good
11. Entropy	High	Low
12. Stability	Poor	Good
13. Information	Low	High
14. Life cycles	Short-simple	Long-complex
15. Bio-chemical diversity	Low	High
16. In organic nutrients	Extra biotic	Intrabiotic

Note : P—R ratio = gross production/community respiration.
P—B ratio = gross production/standing crop biomass.
B—E ratio = biomas supported/unit energy flow.
NCY = Net community yield.

The net result of the complex interaction process of the ecosystem is symbiosis, nutrient conservation, stability, quality, decrease in entropy and an increase in information. The overall strategy is directed towards achieving as large and diverse an orgnic structure as is feasible within the limits set by the available energy inputs and prevailing physical conditions of existence (soil, water, climate etc.). From the above contrast we infer the following characteristics :

Early Stage	Developed Stage
Production	Protection
Quantity	Quality
Growth	Stability
Consumption	Conservation

From the above specifications it is mathematically impossible to obtain a maximum for more than one thing at a time. So one cannot have both extremes at the same time and place since all characteristics are desirable in the aggregate. One appropriate solution entails therefore deliberate planning for compartmentalization of the landscape so as to maintain simultaneously high productive and predominantly protective types of separate units. If ecosystem development theory is valid and applicable to the planning, then the so-called multiuse strategy can be adopted through this approach. More emphasis needs to be placed on compartmentalization so that growth type, steadystate and intermediate type ecosystem can be linked with urban and industrial areas for mutual benefit. Knowing the transfer coefficients that define the flow of energy and the movement of materials and organisms (including the man) between compartments it is possible to determine through analog—computer manipulation, rational limits for the size and capacity of each compartment. One start for example with a simple model consisting of four compartments of equal area, partitioned according to whether the area is a (a) productive, (b) protective, (c) a compromise between (a+b) or (d) urban-industrial. By continually refining the transfer coefficients on the basis of real world situations and by increasing of decreasing the size and capacity of each compartment through computer simulation it is possible to determine objectively the limits that must eventually be imposed on each compartment in order to maintain regional balances in the exchange of vital energy and a materials. Implementation of this compartmentalization plan requires procedures for landscape zoning which does not work very effectively because the zoning restrictions are too easily overturned by short term economic and population pressures. So zoning requires a whole new order of thinking and greater use of legal measures. So preservation of natural areas is not at all a peripheral luxury for the society but a capital investment from which we expect to derive benefits. It may well be that restriction in the use of land and water are our only practicable means of avoiding over-population and too great an exploitation of resources or both. With human rights, environmental rights

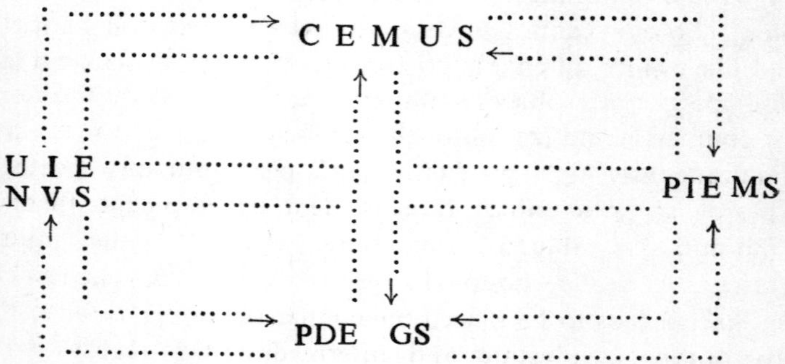

CEMUS—Compromise Environment, Multi-use-system.
UIE NVS—Urban Industrial Environment, non vital system.
PDE GS—Productive Environment Growth system.
PTE MS—Protective Environment, Mature system.

are equally vital. The one man-one-vote idea is accepted but more important is 'one-man-one-acre land' proposition.

Dismal Out Look

As population keep soaring, the emphasis is on production and growth. Most of the regions have passed through early stages of development. They are now in a period of retrogressive succession, brought about by decadence and an impoverishment of the system. The world is becoming standardized. Cities, once distinctive now have a monotonous sameness about them. Now if atoms have become itinerants in the urban-industrial scene, it is because we have a simplified ecosystem. We are destroying high diversity for maximum productivity and efficiency and in doing so we are creating acute problems for ourselves. Respirations tends to be greater than the production. Like an overgrazed pasture the mature system tends to be destroyed and the ecosystem has tended to regress to a stage from which it may never recover.

It is because of the fact that the politico-economic world has not yet concerned in terms of ecosystems and subtle interdependences in nature. Man ceaselessly strives for riches, for growth and for economic development. He is enslaved to and obssessed by technology and the G.N.P. This obsession has despoiled the environment and is ruining the capacity of the ecosystem to support life. The land is scarred and eroded, the waters of rivers, lakes and oceans are so contaminated with industrial wastes, that it is nearly unfit for either industrial use or for human consumption. Air is filled with gaseous and particulate pollutants toxic to life. Pesticides used to promote agricultural production and public health has been severely poisoned the environment. Each agent of production and consumption regard their disposal cost as zero and use the environmental sector as long as it permits him to improve his own welfare. He does not have to pay anything to anybody. Environment is still regarded as a common property, each agent acts as if he owns it. The reckless use without paying any heed to the damage inflicted is causing environmental standard unhealthy and detrimental to urban habitat. Environment poisoning is commulatively reducing the life span and casting lives which is beyond the perception of ordinary people living in the urban-industrial area. So we are coming to a jolting realisation of what have been done to the environment and how we are indeed jeopardizing its future. Obviously the consequences of this vertical growth is ecological disaster.

We have a rude awakening. The inter-action of ecology and economy are being very clear everyday. Ecologically overstressed societies are further impelled to more intensification of pressure on their environment in an effort to survive. For us this is a very dismal outlook, although as the course of events now proceed a true one. With population overcrowding, we can feed them, clothe them and shelter them. But we are going surely to force very many of them to live in niches which are very non-congenial. A review of the past reveals that we have been able to get tremendous amount of energy at relatively low cost. Future costs are going to be much higher however since acquisition of fuel becomes increasingly more expensive as we tend to deplete sources that are easily accessible.* The future therefore holds even greater restrictions on individual freedom. We can never live as our forefathers lived. Posterity is bound to be deprived. Our traditional way of doing things seems like poems of the past.

Looking Ahead

All the while man has been treating the earth as an open system. Natural ecosystem was there to be subdued, to be exploited even to be enjoyed. He was an intruder. But he has infact been operating within a closed system which has been constantly short-circuiting. The earth and its resources,

---

* Even the reduction of pollution is also becoming costilier. Elimination of the erosion of sulfer dioxide and other pollutants from the exhausts of industrial plants costs enormous for every installation. Such costs are recovered only by shifting along to the consumers of the products, produced by that plant.

which have been regarded as infinite is now being discovered as finite. And if man has to exist in a finite system, then he has to work with nature's rule under which he has evolved. Therefore, we have to operate within a complex of dynamic interrelated systems of which man is an integral part. The urgent need is to develop an 'ecosystem approach' to the management of resources. Before making the regional policies first the potentialities of natural ecosystem need to be determined. Ultimately an ecosystem approach tends to evolve a change from a consumptive economy to a conservation-economy oriented towards conservation, maintenance of capital stock and recycling of materials. It can only ensure a sound and substanable balance between a productive environment and a protective environment.

Ecosystem approach is not anti-life. Its fundamental objective is total wellbeing of the society now and in the future. The welfare of the individual is ultimately dependent upon the viability of the life-supporting system. Impoverishment of an ecosystem means inpoverishment of the entire society dependent upon it. Thus ecosystem approach is a compassionate concern for the posterity and for the planet as a whole. Mere economic growth is not enough in bringing welfare to the mankind. Man is an organic being, whose total development depends to a large extent on the non-economic sociological, psychological factors and on a meaningful sustainable balance between him and his environment in which he lives. There should be an organic growth of the entire environment. The man is only a part of the whole. Therefore, an inter-disciplinary approach or holistic approach in planning is sine qua non for the sustainable development of the economy. Environment and growth are not contradictory or incompatible. Growth is rather a pre-requisite infact for "sustainable development" which is defined as "the progress which seeks to meet the needs and aspirations of the present without compromising the ability of future generation to meet their needs. Sustainable development without any economic disasters, something which neither the environment zealots nor those with development facts even though can progress without contradiction.

Sustainable growth is expected to create the capacity to solve environmental problem. So environmental protection and economic development are two sides of the same coin. Development cannot subsist upon a deteriorating environmental resource base. The environment cannot be protected when growth does not take into account the costs of environmental destruction. Therefore development processes should be less energy-intensive and more sensitive to environmental impacts. No country no economic agent should exploit its comparative advantage in "pollution-intensive goods". So we should deal with environmental and developmental issues together. Our common future now has served a notice that the time has come for a marriage of the economy and ecology, so that government and the people can take responsibility not just for environmental damages but for the policies that cause the damages.

One problem-one solution approach is no longer appropriate and must be replaced by an ecosystem analysis that considers man as a part of not apart from the environment. The present scale of ecological research is quite incomplete, inadequate and is still too individualistic to be an effective counter to the lack of balance in the ecosystem. There is thus a great need of studies which can bring about a substantial reorientation of the goals of the society. A solution to the problem of environmental degradation can be achieved through a moral and legal means of mutual coercion, mutually agreed upon by the majority of the people. Permit system, effluent charges, receptors levels, assessment and cost internalising procedures as economic incentives are required for achieving mutually agreed upon coercion.

In a pioneer ecosystem high birth rate, rapid growth rate, high productivity and profit, greater exploitation of accessible and unused resources are advantageous but as the saturation level approaches, these drives must be shifted to the consideration of symbiosis, birth control, and the recycling of resources. A balance between the youth and maturity in the socio-environmental system therefore should be the basic goal, if man as a species is to suceessfully pass through the present rapid growth stage to which he is well-adapted to the eventual equilibrium density stage of which he as yet has little understanding and to which he shows little tendency to adapt. So efforts are essential if we are to maintain any resemblance of the standard of living that has been attained already. We have to reconcile our demand with the need (ecological imperatives) according to a well order plan but not by panicky of impoverishment in the face of imminent shortages. What we need to do is to plan against the increase in population, increase in G.N.P., unlimited dysfunctional growth of cities, use of prime agricultural land for urban development and to plan for new cities, and rebirth of smaller towns and preservation of irreplaceables. We have to place the maintenance and preservation of the fragile ecosystem ahead of technology, population and growth of G.N.P.

References

Bates Martson (1960) The *Forest and the Sea*, Random House, New York.

Breese, Gerald (1978) *Urbanisation in Newly* Developing *Countries*.

Dasmann, R (1968) A *different kind of* country Macmillian, New York.

Donella H. Meadows, et. al.-(1972) The *Limits to growth* Universe Books for Potomal Associates, New York.

Dubos Rene (1968) *Human an Animal*, Scribner, New York.

Gadgil, M. (1985) Towards an Ecological History of India; *Economic and Political Weekly*, Vol. XX, No. 45-47 special No. pp. 1909-18.

Hardin, G. (1968) *Science*, 162 : p. 1243.

Murphy, E.P. (1967). *Governing Nature*, Quadrangle Books, Chicago.

Odum, E.P. (1976) The strategy of Ecosystem development in the *Ecology of Man* (ed) Robert Leo Smith, Section-III, Harper and Row.

Ridkar, R.K. (1973) To grow or Not to grow—thats not the Relevant Question, *Science*, 182, pp. 1315-18.

Smith, R.L. (1976) (ed). The *Ecology of Man* an *Ecosystem* Approach, Harpar & Row.

Tansley, A.G. (1935) The use and Abuse of Vegetational Concepts and terms, *Ecology* 16 : 284-307.

2

Ecological Impacts of Development

*Dr. Peter Wathern**

Introduction

Preventive environmental policy seeks to replace old "reactive", detect and cure, practices with "proactive", predict and prevent, approaches (Simonis, 1988). At present environmental impact assessment (EIA) is one of the few components of preventive environmental policy which is firmly in place and functioning (Wathern, 1988). The value of EIA, in theory, lies in the ability to predict future conditions consequent upon a course of action so that measures can be introduced to prevent them from occurring or to mitigate them. This may involve decisions not to follow a particular course of action or to proceed with built in safeguards.

Ecologists have an important role in EIA, indeed in the past assessment for many developments has often been seen in largely ecological terms. The influence that ecologists are able to exert in order to realise the objectives of preventive

*Professor, University of Wales, U.K.

policy reflects mainly their ability to make predictions about the future state of ecosystems. The basic knowledge upon which these predictions are based is, for most systems, imperfect, although the main deficiencies are slowly being remedied. At this juncture, universal prescriptions for ecological impact assessment are probably undesirable and certainly unrealistic. An understanding of the ways in which ecological systems function and respond to change is of far greater value. This is the basic tenet of this paper.

Thus, ecological concepts are discussed in the first section. In the second section, development is reinterpreted in terms of ecological stress and the way in which individual organisms, populations and systems respond to stress is reviewed in the subsequent three sections. The fifth section comprises guidance on predicting system changes, while the final section contains a review of ecological significance within the decision-making process.

Important Ecological Concepts Relevant to EIA

Ecological systems are characterised by a variety of interrelationships between components including dependency, competition and mutualism. Ecosystems, the basic conceptual unit of ecological study, comprise communities of plants and animals which interact not only within and between these biotic components but also with the abiotic physical and chemical components of the environment. Assessment of the ecological impacts of development is often difficult either because the complexities of these interactions are not adequately known or because ecosystems sometimes react in unexpected ways. Indeed, most of the unexpected impacts of development result from a failure to anticipate effects which are distant, in spatial, temporal or ecological terms from the point of action. In this sense "ecologically distant" implies components which have few obvious linkages between them.

Many ecological impacts are readily apparent and require little skill in predicting. Site clearance results in the destruction of on-site vegetation. This is an example of a direct impact. However, if the vegetation is *Calluna* heath and the site is located in Dorset or Hampshire, a population of the rare Dartford warbler (*Sylvia undata*) may be affected. Even though no birds may be destroyed by the heavy machinery, that is no direct impact, the habitat may be so modified that it will not support this species. This is an example of an indirect impact. It may involve a single dependency change—a second order impact—or may operate through two or more stages in the chain of dependency—third and higher order impacts (see Wathern, 1984). Indirect, especially higher order, impacts are the most difficult to predict. The important dependencies through which indirect impacts are manifest involve energy and materials transfer.

Ecosystems are powered by the transfer of energy, generally in the form of chemical energy (organic matter) between different components. Most are powered ultimately by light energy fixed by plants containing photosynthetically active pigments, although some ecosystems are based upon chemical energy sources. Much of the energy transfer involves consumption of living plants and animals, but detrital feeders utilising non-living organic matter such as faeces and litter are essential components of most systems. The energetics of ecosystems can be described not only in terms of individual organisms or species, but in functional terms as trophic levels. The lowest level in the system are the primary producers, generally green plants, which comprise the first trophic level. Green plants are consumed by herbivores (second trophic level) which in turn are prey to carnivores (third trophic level). In some systems a top carnivore (fourth trophic level) may be present. The food chain which this describes may comprise only four species, a simple and very unusual system, but in general trophic structures are more complex with food webs involving many components.

An important consequence of energy transfer as organic matter is that materials are also transferred. Many are necessary for the functioning of the organism but, increasingly, man-made environmental contaminants are also transferred.

In general, as materials are transferred from one trophic level to the next, the recipient organism is relatively effective in conserving them and may even accumulate them. Shows the importance of bioaccumulation of DDT in an aquatic system. Here there is a 10^7 fold increase in concentration in Ospreys (*Pandion haliaetus*) compared with ambient concentrations. Bioaccumulation of organohalides such as DDT in peregrine falcons (*Falco peregrinus*) in the past has been sufficient to affect reproduction adversely (Ratcliffe, 1970).

The example of bioaccumulation of mercury discharged into Minimata Bay, Japan, is a well known instance where the top trophic level, man, accumulated the metal to lethal concentrations. The interaction between the biotic and abiotic components of the environment in determining effects is also well illustrated by mercury. Inorganic mercury has a relatively low toxicity in comparison with organic, epecially methyl, mercury. For some time it was believed that methyl mercury was purely a man-made chemical. However, it has been detected in water courses where only inorganic salts have been discharged. Microorganisms in the streambed sediments have been found to be capable of methylation into the more toxic form. Thus, bioaccumulation is exacerbated by biotransformation.

Species which make up a particular community are often found, in the presence of other species, to occur over a narrower range of environmental conditions than it is capable of occupying when alone. With respect to a particular parameter, a species will show a bell shaped distribution of numbers against a range of parameter values, with a clearly defined optimum. Of course in reality, the parameter is but one of a large number of components that make up the environment. The species will show a set range of tolerance with respect to each of these parameters. The environmental distribution with respect to all parameters is described as the niche of a species. The extent of the niche that a species is capable of occupying when other species are absent (the theoretical niche) is generally broader than the one occupied when other species are present (the realised niche). Other species may be more efficient at exploiting resources over part of the range of the theoretical niche and confine the species competitively to that part of the niche that it can occupy most effectively. Development may affect the relative efficiency with which competing organisms utilise resources and determine the outcome or redetermine the balance between them.

An advance in ecological understanding which may yet play some role in impact assessment is the recognition of functional groups of organisms within ecosystems. These are described as guilds. Advocates of the use of the guilds point to the advantages of considering groups of species which utilise the same resource in more or less the same way, although occupying different niches (Landres, 1983). The data are more simple, should be easier to collect, allow for shifts in the balance between species over their range and facilitate prediction. Variations in the response of different species within the guild which might complicate predictions can be subsumed within the aggregate response.

One major criticism of guilds is that they lack any rigid definition. This may account for the attractiveness of the concept for some ecologists. Thus, for example, recognition of the aggregate response of the guild to insectivorous birds feeding on broadleaved woodland in South East England may be an appropriate level of analysis for a particular impact prediction. This may restrict its utility. As the guild has only a functional identity attempts to disaggregate the response in order to make statements about members of the guild may be invalid. Similarly, it may be invalid to transfer knowledge about a particular guild to investigations of other guilds which differ significantly. As well as guild definition being subjective, ecologists might find it difficult to agree on guild membership.

Ecological processes can be investigated at various levels which can, in effect, be regarded as a hierarchy of increasing complexity. Ultimately, any change in biota at an ecosystem level is a reflection of the aggregate responses of individual organisms to perturbations. Rarely, however, are there adequate resources within ecological investigations to assess impacts upon individuals even in those situations where "individuals" are clearly delineated. Analysis is more typically centred upon : populations—a group of individuals of the same species within a defined time-space framework; communities—the interacting plant and animal populations within a particular area at a specified time; ecosystems—the interaction of these plant and animal communities with each other and with their abiotic environment. These definitions clearly lack rigour, but most ecologists would accept sufficient elements within them to provide a working system, even though there may be certain "fuzziness" in delineating the tiers in the hierarchy and more specifically great difficulty in applying one of these labels in a particular situation. Furthermore, whether this progressive aggregation results in systems (communities and ecosystems) which are more than the sum of the components, is one of the great current debates in community ecology which, if it is ever resolved, might provide a theoretical basis rather than a pragmatic justification for a particular level of analysis.

Ecological Stress

A generalised theory of ecology and development is still a long way off with the result that other tenets within the overall discipline of ecology must, for the time being, provide guidance on the prediction of ecological impacts. One of the most promising lines of argument seems to be to regard development as an environmental perturbation which places stress or stresses upon ecological systems. There have been various attempts to refine an acceptable definition of stress in biological terms (see Ivanovici and Wiebe, 1981), but with little concensus. For the purposes of this discussion, stress as a developmental phenomenon will be defined (modifying the views of Bayne, 1975 and Barrett et. al., 1976) as a perturbation applied to a system which is either foreign to that system or, if natural, is applied at an excessive level which renders that individual, population, community or ecosystem vulnerable to further change. It may be manifest at a variety of levels in the ecological hierarchy.

Effects of Stress on Individuals

Most of our understanding of the effects of stress events upon individuals comes from a study of humans, with some support from investigations of laboratory animals, but only infrequent—often casual observations—related to individual plants and animals in field situations. Practical difficulties in field research, not least resource limitations, severely restrict the ability to study the response of individual plants and animals to change. Furthermore, ecologists tend to be interested in stochastic processes which are more amenable to investigation at the population level and above. In marked contrast, our perception of people is very much in individual terms with the result that a great deal of evidence on the effects of stress at this level comes from human medical and psychological research.

A variety of chemical and physical factors induce change in individuals. In some cases, epidemiological studies indicate correlations, while the exact causal mechanism remains to be elucidated. Ultimately, however, these effects have a biochemical basis which is often not known. That there is considerable inter-relationship between physiological, psychological and behavioural impacts which should perhaps be regarded as a complex of effects.

The interaction of the environment and the individual has long been known to influence the outcome of potential stress events. With detailed statistical records it has been possible in many developed countries to identify genetic predispositions to certain medical conditions which have a proximal cause which is environmental. Whether a similar situation applies in other species would be impossible to assess from field investigations, but many investigations underline the importance of this genetic effect in laboratory animals.

Not only are individuals at heightened risk, but certain

biological processes are particularly vulnerable to stress. Reproduction may be impaired at even low levels of stress. Thus, in human populations wartime experiences indicate that fertility is affected by an insufficient or inadequate diet. In domestic animals lactation and the birth of young have critical thresholds of dietary quality, for example with respect to phosphate levels. Even if birth occurs the young may be adversely affected by the parental environment, either prior to or during pregnancy, such as exposure to ionizing radiation or mutagenic chemicals which may cause genetic abnormalities. The environment is also an influence on developing young and may cause teratogenic effects, for example by exposure to drugs or certain chemicals during pregnancy. For young developing independently in the environment, such as shellfish larvae, teratogenic effects may also occur. Radiation and certain chemicals in the environment may also have carcinogenic effects. Ultimately, severe stress events will cause the death of individuals.

Effects of Stress on Populations

Populations are the lowest level in the hierarchy in routinely investigated in field situations; the concern is to elucidate the aggregate response of a number of individuals of the same species. These reponses are summarised in Table 2. Only in special situations are natural populations composed of genetically identical individuals (genotypes). This occurs when populations are derived from a single individual by vegetative spread or other asexal means or are formed within apomictic breeding systems. More usually, natural populations are inherently viable with a range of different genotypes represented. This genetic diversity means that a population has the potential to respond to a range of environmental changes within the niche limitations of that species. One particular genotype or groups of genotypes is likely to be favoured by specific environmental conditions, and the population will persist, with its genetic composition drifting over time in response to environmental change. This maintenance of genetic "options" is believed to confer stability over time in the face of random environmental fluctuations.

Close to the boundaries of the niche of a particular species the environment becomes progressively sub optimal. As a consequence, resources may not be utilised effectively with the result that a more successful competitor may eliminate the species over this particular portion of its potential range. The realised niche, that is the ecological space which is actually occupied, is generally more restricted than its potential ecological range, generally referred to as the theoretical niche.

Environmental stress affects population numbers both directly and indirectly, a phenomenon which is fully exploited in the use of biological indicator organisms to monitor environmental change. The progressive elimination of sensitive lichen and bryophyte species from areas with polluted air is so well documented that biological scales for assessing the levels fo air pollution at a particular locality on the basis of its bryophyte and lichen assemblages have been developed. The elimination of these species occurs because the induced change applies a stress lying outside their respective niche limits. Not all organisms are directly affected by change nor are the changes necessarily adverse. Stress may affect competitive relationships within communities and shift the balance subtly in favour of one competitor which will be able to expand its population numbers or extend the range. Recording the spread of exploiter organisms, such as the air pollution tolerant lichens or *Enteromorpha* spp. in polluted estuaries utilises this phenomenon in environmental monitoring.

One consequence of severe environmental stress is limited genetic diversity as a result of small population size. Small populations are likely to be particularly vulnerable to change as they lack the genetic resilience conferred by genetic diversity.

In some instances, the perturbation sets very narrow constraints upon the range of ecotypic variation that a locality can support, with the elimination of all organisms outside of these precise genetic limits. As a consequence, uniform populations develop which lack the ability to respond to further change. For example, within two or three decades of abandonment the heavy metal contaminated mine waste of mid Wales started to develop a vegetation cover. These individuals, predominantly grasses, were subsequently found to be metal tolerant genotypes. These populations have developed by the rigorous selection for toxitolerant genes naturally occuring within populations at a frequency of about 10^{-4}. Seed derived initially from these populations is now used commercially to reinstate metal contaminated sites. The cost of metal tolerance to the individual grasses is high, for part of the toxitolerant strategy is an inherently low growth rate, with the result that toxitolerant individuals are rapidly eliminated by intraspecific competition in metal free situations. The ecological consequences of genetically uniform populations are evident to agriculturalists who are aware of the implications for disease and pest resistance of genetically identical crops. Genetic diversity has been sacrificed in order to maximise short-term yield.

Finally, populations show variation in the state of development of their constituent organisms, either on a spatial or a temporal basis. A population developing as a cohort, that is with all individuals at the same stage in their life cycle, will be at risk of extinction if catastrophic events coincide with vulnerable phases. Not all stages of the life cycle may be affected equally by environmental stress. For example, the well known effect of atmospheric pollution on saxicolous lichens follows a progress sequence of impact affecting successively establishment of new colonies; reproduction by established colonies; death of mature thalli. Thus, populations of lichens on old gravestones in urban centres lack young recruits because of the failure to establish. At progressively higher pollution levels, the individuals are sterile, lacking reproductive structure; finally they become moribund and are then eliminated.

Conversely, a mixed population comprising many developmental stages may be perpetually at risk from the effects of stress. The systematic elimination of a certain developmental stage, for example young, may lead to a progressively ageing population with a consequent loss of reproductive potential for recovery should the stress subsequently decline. As absolute numbers may remain static for prolonged periods in long-lived species, the adverse effects of stress may be masked because of an inability to determine either the population's age structure or the rate of recruitment of juveniles.

Effects of Stress on Systems

The effects of stress may be experienced at a community or ecosystem level. Such system impacts are the major concern of ecologists involved in EIA. These can be grouped into three types of stress which affect : ecosystem energetics; materials cycling; and structure complexity as a result of simplication (Table-3).

The most obvious impact upon energetics comes through the removal of organisms. For example, constructional activity will generally result in the destruction of much of the vegetation of site with a consequent loss of photosynthetic capacity. Indirectly, development may affect the energy inputs into systems by intercepting radiation. Increased turbidity in water bodies as a result of sediments from construction sites may reduce primary production, a situation paralled in terrestrial systems by, for example, dust deposition on vegetation or by particulates in the atmosphere.

Thermal parameters may be important in determining the niche limits of species, for example in aquatic systems, and a change in the thermal loading of water bodies may be a major influence of development on such systems. Thus, the impact of cooling water discharged from power stations or industrial plant may affect not only species directly by elevated water temperature, but also have indirect effects upon the species with which they compete or which are dependent upon them as a food source.

One consequence of these direct and indirect effects may

be a change in the trophic structure of food chains and food webs, particularly if key species or key relationships are affected. The harvesting of whales in Antarctic waters has probably affected other components of the ocean ecosystem which utilise krill-based systems. The rapid recovery of fur seal populations during the century after the cessation of hunting may reflect an indirect affect of the relaxation of pressure on krill consumption. That fur seals do not feed directly upon krill, but utilise other organisms which are in turn dependent on krill, illustrates the complexity of the ecosystem response to stress which may be manifest through a series of intermediaries.

It was Holling (1978), however, who pointed out that not every element in a system is dependent upon everything else. The effects of a perturbation within a subsystem may be dissipated through compensatory change within that sub-unit, leaving the remainder of the system unaffected or little affected. Thus, increased fur seal numbers may be envisaged as part of the ecosystem response to the reduction in whale numbers. What the effect on the biota of the Antarctic Ocean of krill harvesting by man will be remains to be seen. In general, the resilience of an ecosystem to accommodate or recover from stress appears to be related to the point of perturbation. Impact seems to be greatest when the change affects primary production or involves one of the initial trophic levels. Virtual elimination of many whale species has not brought about the total collapse of the system which appears to have great resilience to this particular change. The implications for conservation, however, is a major concern discussed in detail later.

Measuring energetic status has presented methodological problems. It has been important to devise an objective measure in order to demonstrate the chronic effects of stress which are often not readily exhibited, yet which predispose an ecosystem to further change. In contrast, acute impacts may be readily apparent because of changes in energy flow or in number of organisms. At chronic levels impaired or stressed physiological function can be determined by measuring the adenylate energy change (AEC). AEC is a measure of the extent of phosphorylation of the high energy phosphate molecules which power biochemical reactions. Values range between 0 and 1 with high levels indicative of a high metabolic potential (Atkinson, 1977). In unstressed systems, the AEC of organisms is typically of the order 0.8-0.9. Under conditions of stress, however, AEC values drop below this to between 0.5-0.7 (Ivanovici and Wiebe, 1981).

Man's activities can also affect materials and nutrient cycling. The classic experimental forest clearance in the Hubbard Brook catchment study has clearly demonstrated how development (in this case clear felling as in logging) disrupts nutrient cycling. In the year after clear felling the catchment lost 52.8 kg of nitrate per hectare compared with unaffected controls (Likens et al., 1970). Even apparently benign activities can affect cycles. Traditional forestry practices have had an effect on the nutrient depletion of British woodlands by the continued removal of forest produce through historic times, even under conditions of continuous woodland cover.

A more usual consequence of development is to raise the nutrient budgets of systems. For example, agricultural intensification causes eutrophication not only of aquatic systems by surface washout of fertilisers or slurry into water courses or to ground-water, but also of terrestrial systems by fertiliser drift. The discharge or nutrient rich effluent, such as untreated sewerage, will also cause eutrophication.

Man's record in the use of natural systems, for whatever purpose, is a history of simplification. The replacement of diverse natural systems by genetically uniform crops provides an obvious example. The diversity of systems in terms of numbers of species, so-called α diversity, is greatly reduced. The spatial heterogeneity within systems which dictates the scale of pattern of communities within a system, β-diversity, is also affected as uniform environmental conditions are created over extensive areas.

Some of the changes are intentional. For example, recent agricultural intensification in the karst landscape of the Burren in Western Ireland has been concentrated selectively on the dwarf hazel woodlands. It is clear that the farmers are using woodland as an indicator of moderately deep soils which will carry an, albeit poor, grass crop once the trees have been bulldozed out. The species rich woodlands, carrying some of the most diverse (and rarest) bryophyte and lichen floras in Western Europe, are replaced by agricultural grasslands of no more than four or five different species per square metre. In agricultural grasslands, simplification is also a function of increased nutrient status.

Other components of ecosystems are lost unintentionally often as a result of a lack of basic ecological knowledge about how systems function. Afforestation in Northern England and Southern Scotland has led to the loss of significant numbers of ravens (*Corvus corax*) even though their nest sites have remained intact (Marquiss et. al., 1978). Land-use change appears to affect food resources and the same may be true of the much rarer red kite (*Milvus milvus*) in Wales, but the evidence is not unequivocal (Newton et. al., 1981).

The final purported effect of stress upon systems (Table 3) is a reduction in stability as a consequence of simplification with ecosystem collapse as a possible final outcome. The debate concerning diversity and stability, which underlies this contention, is another of the great debates of community ecology. Advocates argue that diverse systems are inherently more stable than simple ones because of the ability to respond to fluctuations by the transient expansion in range or numbers of some of the elements which, once the system returns to equilibrium, fall back to their former distribution. If this argument is valid, simplifying a system by eliminating species will affect its ability to respond to change and hence influence its long-term stability. Ultimately, it will be unable to respond to stress and cease functioning. Not all ecologists accept that more diverse systems are inherently more stable than simple ones.

Predicting Population and Ecosystem Impacts

From a review of a number of Canadian impact assessments, Beanlands and Duinker (1983) concluded that, if they are made at all, "... predictions are more apt to be vague generalisations about the likelihood of certain conditions prevailing during and after project construction, as opposed to quantified forecasts which could be subject to verification". This reflects, to part, methodological problems and partly a tendency on the part of many ecologists involved in impact assessment to make "academic" pronouncements which abound with qualifying caveats.

The methodological problem reflects the fact that many existing models of environmental change are too coarse-grained for specific predictions of impact to be made in particular situations (see, for example, Hecky et al., 1984). When ecosystem function is well established and the characteristics of the development are known, impact prediction can often be modelled accurately. Sadler (1986) points out that, in Canada the "reservoir paradigm" for hydroelectric development projects is so well established that impacts can be quantified and assessment has become almost routine. For other situations where ecosystem function is poorly understood, assessment may be difficult, especially if a novel type of development it involved. This situation should improve, however, as the results of basic ecological research are used to refine conceptual and mathematical models of ecosystem function.

After nearly two decades of EIA, it is disappointing that many ecologists are still not prepared to make the definite statements about impacts that are necessary for decision making. This unwillingness to back professional judgement means that ecological aspects will not be given due consideration. The development of this capacity is part of the learning process of any ecologist involved in EIA.

Even though in some respects each development proposal will have certain peculiarities which render a definitive ecological impact assessment checklist a pipedream, certain generalities exist to make general guidance feasible. Ecologists carrying out an impact assessment should seek to resolve a number of basic issues.

There should be adequate knowledge about the functioning of the ecosystem under consideration to identify the key components, species and relationships, within it. This need not imply an exhaustive field research programme, for appropriate information may be gleaned from a variety of secondary sources. The possibility of further research as part of a baseline study, however, cannot be ruled out. In conjunction with this, it is necessary to determine whether any of the key elements would be affected by the proposed development and adequate information on the project is clearly a pre-requisite. It is not an efficient use of resources to investigate any elements, even key components, which are not likely to be affected. The concept of key components is discussed more fully in the next section.

It may be possible to predict and subsequently to detect impacts upon key components directly. This ability will depend not only upon a thorough appreciation of the processes involved, but also an understanding of the natural variability within the system. Often an important function of baseline studies is to aid quantification of the variability. Holling (1978) defined "envalopes" of variability within which an environmental parameter or ecosystem component may oscillate and still be stable. Only when it is pushed outside of these statistical confines can impacts be said to occur. In many instances variability is so great that only major deviations are indicative of an impact unless very intensive detailed sampling programmes are adopted when smaller changes can be shown to be statistically significant (Table 4). This clearly underlines the need to make ecological aspects of impact assessment a quantitative exercise.

When impacts cannot be investigated directly some indirect or surrogate measures may have to be determined. First, indirect assessment may exploit the position of a particular organism or key relationship in the trophic structure of the system. Thus, there may be practical difficulties in determining the absolute population of a particular species and it may be equally valid to obtain a relative assessment through counts of prey or predator organisms associated with it. At Point Barrow in Alaska, the number of breeding predatory birds, jaegers (skuas) and owls, is dependent upon lemming numbers (Pitelka et al., 1955). The predatory birds are easier to census than the small mammals which may require detailed "capture-recapture" studies in order to estimate population size.

Secondly, it may be necessary, or difficult to monitor populations throughout their life cycle or as adults. Assessment may give more insight if immature or sedentary phases are studied, particularly if these are particularly sensitive to change. Harris, for example, has shown the difficulties in assessing the effects of environmental change on Puffin (*Fratarcula arctica*) populations. Censuses of breeding performance at fixed locations seem to provide one of the few reliable quantitative measures. Some of the types of effects on individuals described in the previous section (see Table 1) may provide more effective guidance on potential impacts, be more readily predicted as a result of experimental investigations (see below) and be more easily detected after implementation by monitoring.

Frequently, it may be more effective to concentrate upon habitat aspects as, in general, these are easier to predict and monitor. It is important that the habitat components which are selected for assessment are meaningful in ecosystem terms. Thus, it may be appropriate to predict impacts upon the distribution of *Zostera* beds in the intertidal zone, rather than attempt to assess Brent geese (*Branta bernicla*) population changes which may be, in part, regulated by events in their northern summer habitat. The seagrass *Zostera* forms a major food resources during the winter period. Whereas quantitative habitat change may be relatively easy to determine, qualitative change may present greater difficulties. Studies of the distribution of *Sphanga* in raised bogs are relevant for a variety of potential impacts, especially on the hydrology of these systems. However, it may be difficult subsequently to isolate the causal agent of change in complex situations such as this.

Beanlands and Duinker (1983) highlight the use of indicator species in impact assessment. In ecological investigations ecologists will often note the presence of particular species and from its occurrence will often infer a great deal about associated species which may, for example, appear at a different season and about environmental conditions. Indicators are used particularly in baseline studies and in impact monitoring. They provide surrogates for the parameters which are of real concern and are used mainly on considerations of cheapness and convenience. Baker (1976) has described the different ways that they are used in impact monitoring.

There are two additional approaches to impact assessment which have yet to realise their full potential, namely experimentation and simulation. Ecology has moved beyond the early stages of scientific development, namely observation, description and analysis, and progressed to an experimental science. As yet, there is little evidence of experimentation in impact assessment. The major exception to this is toxicity testing for potential pollutants and novel chemicals. Laboratory LD_{50} tests are in effect experiments on animal populations to determine the lethal dose (in LC_{50} the lethal concentration) which will kill fifty percent of the population. Similar field trials are carried out on a range of products during testing prior to marketing. The use of Lund tubes to investigate the effects of nutrient enrichment on aquatic systems provides a some realistic field investigation. The use of pilot-scale experiments in impact assessment is discussed in ward (1978).

Holling (1978) has advocated the use of simulation modelling in impact assessment, indeed has proposed an approach to EIA based upon "sound ecological principles" called adaptive environmental assessment and management. One of the key elements of this approach is the use of computer models to predict the outcome of development actions. A model of the system, or subsystem, likely to be affected is produced, based upon an understanding of ecosystem function. The real value of computer-based systems lies in the ability to run repeated simulations involving variations in the assumptions and mitigating measures built into the model. In addition, the requirement to express ecosystem functions in the form of a definite algorithm highlights areas where current knowledge is inadequate. Although used to only a limited extent in EIA to date (see Bisset, 1988), there is clearly much potential in the use of simulation models in EIA.

Ecological Significance and Decision Making

Whaling is almost universally condemned as a "bad thing", yet in terms of impact upon the Antarctic Ocean ecosystem it is, at worst, an "indifferent thing" because of the redirected energy flow that has ocurred. This difference in evaluation is simply a reflection of the perception by the public in most countries that whales are species worthy of conservation. This translates pretty directly into political action. In assessing ecological impacts, the concern of the ecologist is to improve the quality of information upon which decisions are based, for as Truett (1978) pointedly states "...lest we forget, the ultimate purpose of an assessment study is to influence decisions". The other important message that Truett highlights is that species and ecosystems "......with little value are of little concern to decision-makers".

Beanlands (1988) has discussed the long-standing attempts to define the focus of an impact assessment through scoping. As the objective of scoping is to identify the information which must be obtained during EIA, scoping issues affect ecological information requirements and, hence, its use in decision making. With respect to biological and ecological aspects of impact assessment it is evident that complete and detailed analysis of whole ecosystems are both impossible and unnecessary within the timescale of EIA for most projects. Only important, so-called key or cornerstone, components and relationships can be considered. Almost since the inception of EIA attempts have been made to arrive at a workable definition of key elements.

The US Atomic Energy Commission (USAEC) defined an animal or plant species as important"...(1) if it is commercially or recreationally valuable, (2) if it is rare or endangered, or (3) if it affects the well-being of some important species within criterial (1) or (2) above, or (4) if it is critical to the structure and function of the ecological system". (USAEC, 1973). In investigations of ecological impact, comparable community and ecosystem criteria should also be employed. Thus, EIA should focus on the "...components of the environment for which there is public (including professional) concern". (Beanlands and Duinker, 1983). Furthermore, when ecological impacts have been determined they must be assessed for significance with respect to these four criteria.

There are, however, practical difficulties in adopting some of the USAEC criteria in assessing ecological impacts, either in scoping (as Beanlands (1988) discusses) or to evaluate the output from ecological investigations which is the concern here. Economic or recreational importance should be relatively uncontentious. Important methodological advances have been made recently in determining the relative significance of systems and species affected by development. On the one hand, a range of statutory designations provide an established measure of excellence for elite sites. In the U.K., for example, these include designation as Ramsar site, National Nature Reserve, Special Protection Area, Site of Special Scientific Interest, World Heritage Site. Similarly, red data books provide information on rare and endangered species. In many countries, and for some groups of organisms, asssssment of the ecological significance of other sites and species has gone some way towards resolution. Ornithologists use the 1% criterion to argue international, national or regional significance for a site with respect to an individual species if it contains, respectively, more than 1% of the total of that particular population (Saejis and Baptist, 1977). In the Netherlands, sites can be compared with a standard value for the average diversity of a unit area of the country as a whole (van der Maarel, 1970). Elsewhere, criteria have been developed for determining rare species (Adamms and Clough, 1978) and a wide range of so-called "ecological evaluation" techniques have been produced for assessing the relative importance of sites for nature conservation purposes (see, for example, the discussion in Wathern et al., 1985).

Grave difficulties remain over defining species which either affect commercially or recreationally valuable or rare or endangered species (criterion 3) or which are critical to the structure and functioning of a particular ecosystem (criterion 4). Such pronouncements are dependent upon basic understanding of ecosystem function and these will only improve in line with fundamental ecological research on particular ecosystems. Where past performance in carrying out ecological impact assessment has been poor this has reflected primarily a lack of ecological knowledge. This is not a problem confined to ecological investigations for Culhane (1987) has reported from a study of a number of EISs produced in the US, that the assessment of social and economic impacts is frequently far worse. Again inaccuracy is a consequence of a dearth of knowledge on system function.

A great deal of generalised information on ecosystem processes, in particular energy and materials transfer, has been provided by the International Biological Programme (IBP). This has greatly increased understanding of the functioning of many of the world's biomes which provides some pointers for ecological aspects of EIA. These studies enable the major transfer points between components of ecosystems and, hence, key relationships to be identified. What is lacking is an understanding of the variability of many systems which affects the ability to make valid statements about detailed ecological impacts in particular situations.

Conclusions

On the basis of an investigation of EISs produced in the US, Culhane (1987) concluded that EIAs were "...inadequately scientific exercises". He concluded that without collecting systematic data on the impacts of a development, those involved in EIA cannot hope to refine their predictive capabilities. More positively pointedly, he concluded that EIA forecasts are "...not inaccurate". According to his study, few impacts were demonstrably inconsistent with forecasts, even fewer were unexpected and, of these, none was major. Similarly, acceptance that hydroelectric projects in the sparsely populated northern environments of Canada can be predicted consistently (Sadler, 1988) indicates a general competence in understanding ecosystem function. In contrast, Beanlands and Duinker (1983) considered the scientific quality of EIA from an ecological perspective to be a dismal record. Their criticisms were that EIAs : were descriptive not predictive; lacked a rigorous approach to analysis and interpretation; provided results of doubtful value for decision-making; suggested mitigation measures based upon generalised principles rather than specifically orientated to the proposal; and lacked a capability of testing prediction through monitoring.

These conclusions seem to indicate that ecologists involved in EIA still have major deficiencies in working practice to overcome, but that, in general, the present understanding of ecosystem function is adequate for the purposes of EIA in many environments. The major problem that ecologists have to overcome is to move away from generalisation and vagueness. Decision makers are interested in two features of an impact, namely probability and consequence. Even improbably impacts are likely to elicit concern if their consequences are dire, while high probability events of little consequence are likely to be ignored. Uncertainty, therefore, is a primary concern in EIA (de Jongh, 1988); indeed, Holling (1978) has described EIA as the management of uncertainty.

Ecologists will have to ensure that the format of their predictions conforms to the requirements of the decision maker. A primary need is a statement concerning the consequences of change. Ideally, these should be numerical predictions about the likely extent of an impact in terms of, for example, geographical area or population change, within a defined timescale. The second need is some indication of the probability of an impact occurring.

The ability of an ecologist to make such statements depends in part upon a willingness to back reasoned professional judgements. The validity of these judgement depends also upon the current state of knowledge about ecosystem function. Thus, the predictive capability should increase as improvements in basic knowledge enable more valid conceptual and mathematical models of likely change to be produced. With these improved models, simulations should accord more wtth reality and become more useful in EIA.

Inevitably, there will be situations where ecological knowledge is inadequate to make predictions. In these situations, the proper application of EIA procedures may help to improve the ability to deal with the assessment of change which affects key species and key relationships. Adequate monitoring programmes in conjunction with past development audit should reveal deficiencies in prediction and indicate uncxpected ecosystem responses to stress. The implementation of any major development project can, in effect, be a large experiment to investigate ecosystem responses to stress as long as adequate resources are invested in environmental monitoring.

Table 1

Some Examples of Environmental Stress Effects upon Individual Organisms

Type of Effect	*Agent*	*Source*	*Impact*
Physiological	lead	traffic	retards intellectual development of children
psychological	aluminium	acidification of upland soils	Altzheimer's disease (senile dementia) affects mental processes of old people
behavioural	various	food additives	? hyperactivity in children
fertility	reduced phospate or energy intake	poor quality or loss of food	fertility loss when intake falls below thresholds
genetic (mutagenic)	radiation	various	effects on DNA leading to genetic abnormalities in offspring

developmental/birth (teratogenic)	PCBs, Cu, Hg, paraquat etc.	various	failure of organs of foetus to develop properly
	TBTs	antifouling paints	abnormal developments of shellfish larvae
cancer (carcinogenic)	radiation	nuclear power stations	? leukaemia clusters in children
	hydrocarbons	atmospheric pollution	lung cancer
	not known	water pollution	? skin lessions in commercial fish species in North Sea
death	various	various	

? in Tables 1-3 denotes an unproven impact.

Table 2

Some Examples of Environmental Stress Effects upon Populations

Type of Effect	*Agent*	*Saurce*	*Impact*
change in number	macronutrients	sewerage discharge	increase in populations of organisms comprising "sewage fungus"
	heavy metals	smelters	loss of sensitive bryophytes and lichens
population structure	SO2	burning sulphur fuels	change in population structure of saxicolous lichens on gravestones
survival of tolerant genotypes	heavy metals	nine waste	development of toxitolerent vegetation on heavy metal mine wastes
	soot	air pollution	industrial mechanism in pepper moths and ladybirds
simplification of genotypes present in population	various	various	reduced ability of population to adapt to further environmental change or developmental impacts, e.g. low pest resistance in monuculture crops.

Table 3

Some Examples of Environmental Stress Effects Upon Ecosystems

Disruption of Energetics

Type of Effect	*Agent*	*Source*	*Impact*
1	*2*	*3*	*4*
loss of photosynthetic potential	habital loss	construction activity	destruction of plants reduces energy fixation and total amount available to system
changes in energy input	sediments	construction activity	increased turbidity in water courses leading to reduced photosynthesis
	dirt	quarrying	dirt on leaves reduces photosynthesis
change in heat balance	heat	thermal pollution from power station	impact on sensitive organisms

Contd. on page 40)

Table 3 (Contd.)

1	*2*	*3*	*4*
trophic structure	species loss	whaling	modified energy fluxes through different trophic levels
community energetics	various	various	reduced adenylate energy charge
Disruption of Chemical Cycles			
nutrient depletion	macronutrients	forestry	clear felling loss of nutrient buffering capacity
eutrophication	macronutrients	sewage works	previous systems unable to cycle superabundant nutrients
Simplification			
loss of structural complexity	habitat destruction	construction	loss of complex systems generally replaced by simple ones
loss of filled niches	food resource	afforestation	reduced raven and ? kite population
reduced diversity	overgrazing	pocr range management	reduced diversity of communities to grazing tolerant species

reduced diversity	homogeneity	golf green management	uniform seed mixture, fertiliser, irrigation and cutting regimes
simplified food webs	reduced home range	land use change	loss of specialist organisms with large home ranges, e.g. raptors, carnivores
lowered stability ?	various	various	? possible ecosystem collapse

Table 4

Effect of Data Variability on Sampling Requirements

Variability of data expressed as coefficient of variation ($s/\bar{x}$) of raw data	*Approximate number of samples required to ensure 90% confidence that the sample mean will lie within ± the given percentage of the true mean*		
	40%	*30%*	*20%*
0.1	3	3	3
0.2	3	4	5
0.4	5	7	13
0.6	7	13	27
0.8	13	22	46
1.0	19	32	70
1.2	27	46	100
1.4	36	61	135
1.6	46	79	174
1.8	57	100	220
2.0	70	123	271

After Welsh Water Authority (1982)

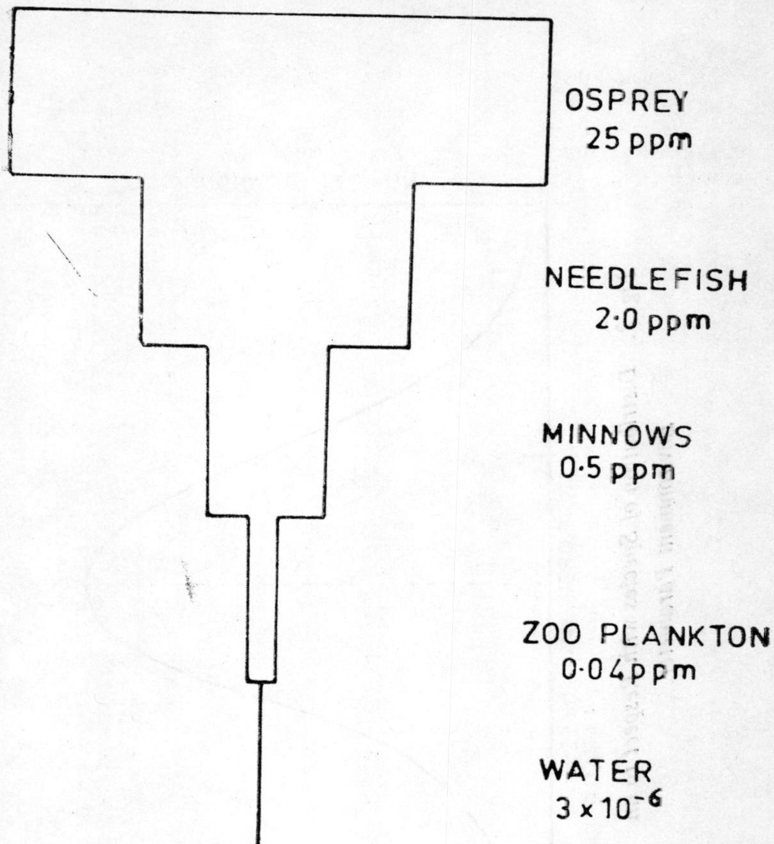

Fig. 1. Bioaccumulation of DDT in an Estuary of the Eastern USA.

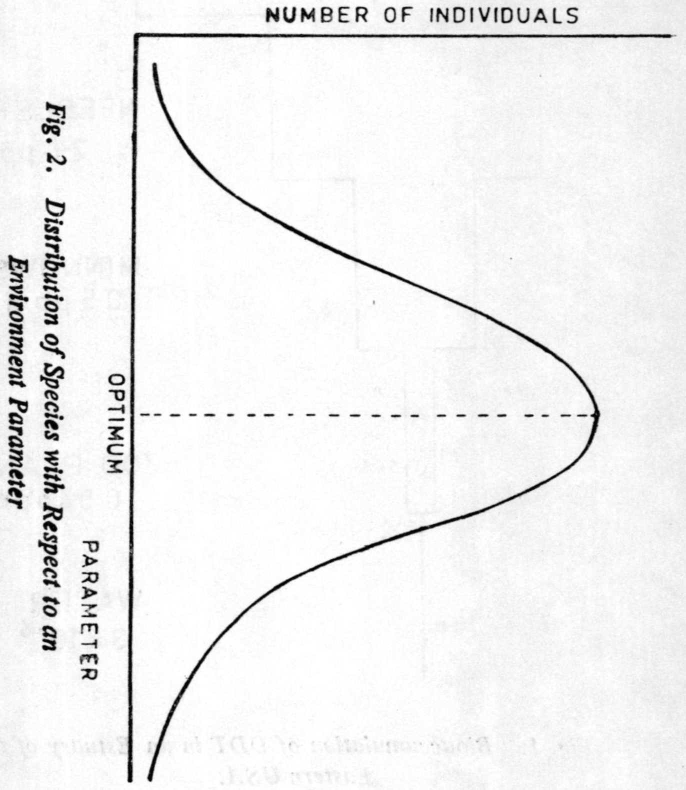

Fig. 2. Distribution of Species with Respect to an Environment Parameter

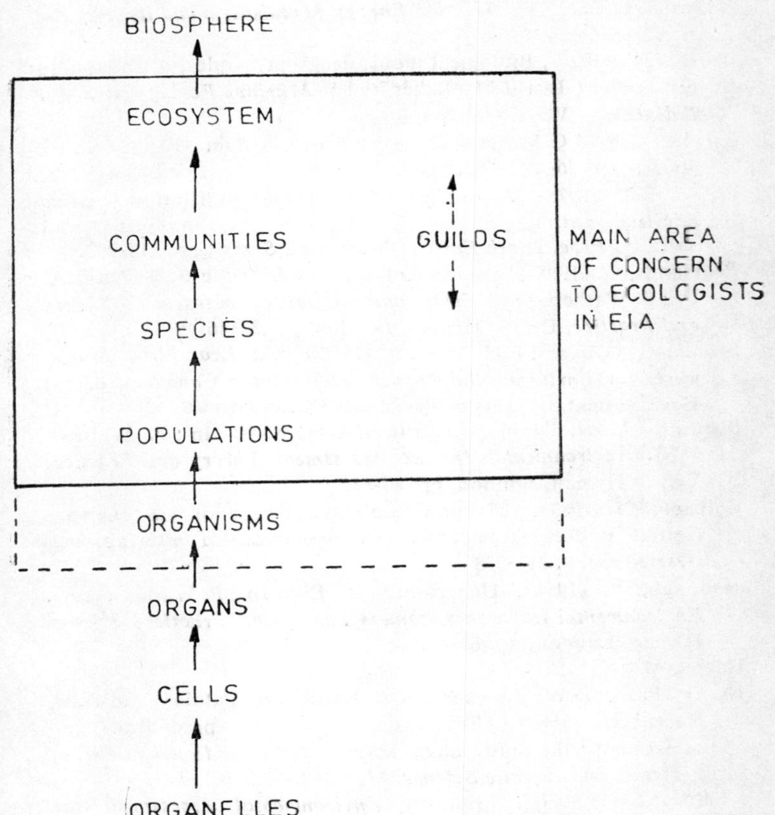

Figure 3. Hierarchy of Ecological Organisation

References

Adamms, P.R. and G.C. Clough (1978), Evaluating species for protection in natural areas, *Biological Conservation*, 13, 165-178.

Atkinson, D.E. (1977), *Sellular Energy Metabolism and its Regulation*, Academic Press, London.

Baker, J.M. (1976), Biological monitoring—principles, methods and difficulties, In : J.M. Baker (ed.) *Marine Ecology and Oil Pollution*, Wiley, Chichester, pp. 41-53.

Barrett, G.W., G.M. van Dyne and E.P. Odum (1970), Stress *Bioscience*, *26*, 192-194.

Bayne, B.L. (1975), Aspects of physiological condition in *Mytilus edulis* L., with respect to the effects of oxygen tension and salinity, *Proc, Ninth Europe, Mar. Biol, Symp.*, pp. 213-238.

Beanlands, G. (1988), Scoping methods and baseline studies in EIA. In : P. Wathern (ed.) *Environmental Impact Assessment Theory and Practice*, Unwin Hyman, London, pp. 33-46.

Beanlands, G.E. and P.N. Duinker (1983), *An Ecological Framework for Environmental Impact Assessment in Canada*, Federal Environmental Assessments Review Office, Ottawa.

Bisset, R. (1988), Developments in EIA methods. In : P. Wathern (ed.), *Environmental Impact Assessment Theory and Practice*, Unwin Hyman, London, pp. 47-61.

Culhane, P.J. (1987), The precision and accuracy of US environmental impact statements, *Environmental Monitoring and Assessment*, 8, 217-238.

de Jongh, P. (1988), Uncertainty in EIA. In : P. Wathern (ed.), *Environmental Impact Assessment Theory and Practice*, Unwin Hyman, London, pp. 62-84.

Harris, M.P.

Hecky, R.E., R.W. Newbury, R.A. Bodacy, K. Patalas and D.M. Rosenberg (1984), Environmental impact prediction and assessment : the South Lake experience, *Canadian Journal of Fisheries and Aquatic Science*, *41*, 579-90.

Holling, C.S. (1978), *Adaptive Environmental Assessment and Monitoring*, Wiley, Chichester.

Ivanovici, A.M. & W.J, Wiebe (1981), Towards a working 'definition' of 'stress' : a review and critique, In : G.W. Barrett and R. Rosenberg (eds.), *Stress Effects on Natural Ecosystems*, Wiley, Chichester, pp. 13-27.

Landres, P.B. (1983), Use of the guild concept in environmental impact assessment, *Environmental Management*, 7, 393-398.

Likens, G.E., F.H. Borman, N.M. Johnson, D.W., Fisher and R.S. Pierce (1970), Effects of forest cutting and herbicide treatment on nutrient budgets in the Hubbard Brook ecosystem in New

Hampshire, *Ecological Monographs*, *40*, 23-47.

Marquiss, M., I. Newton and D.A. Ratcliffe (1978), The decline of the raven (*Corvus Corax*) in relation to afforestation in southern Scotland and nothern England, *Journal of Applied Ecology*, *15*, 129-144.

Newton, I., P.E. Davis and D. Mass (1981), Distribution and breeding of red kites in relation to land use in Wales, *Journal of Applied Ecology*, *18*, 173-186.

Pitelka, F.A., P.Q. Tomich and G.W. Treichel (1955), Ecological relations of jaegers and owls as lemming predators near Barrow, Alaska, *Ecological Monographs*, *25*, 85-117.

Ratcliffe, D.A. (1970). Changes attributable to pesticides in egg breakage frequency and eggshell thickness in some British birds, *Journal of Applied Ecology*, *7*, 67-107.

Sadler, B. (1988), The evaluation of assessment : post-EIS research and process development, In : P. Wathern (ed.), *Environmental Impact Assessment Theory and Practice*, Unwin Hyman, London, pp. 129-142.

Saijs, H.L.F. and H.J. Baptist (1977), Wetland criteria and birds in a changing delta, *Biological Conservation*, *11*, 251-256.

Simonis, U·E. (ed.) (1988), *Preventive Unweltpolitik*, Campus, Frankfurt am Main.

Truett, J.C. (1978). Ecosystem process analysis—a new approach to impact assessment, *Proc. Energy/Environment '78 Symposium*, pp. 70-75.

United States Atomic Energy Commission (1983), *Preparation of Environmental Reports for Nuclear Power Stations*, Directrate of Regulatory Standards Regulatory Guide 4.2, United States Atomic Energy Commission, Washington D.C.

van der Maarel, E. (1970), De Ooijpolder, een biologische waardering van natuur en landschap, *Natuur en Landschap*, *24*, 201-233.

Ward, D.V. (1978), *Biological Environmental Impact Studies, Theory and Methods*, Academic Press, London.

Wathern, P, (1984), Methods for assessing indirect impacts. In : B.D. Clark, A. Gilad, R. Bisset and P. Tomlinson (eds.), *Perspectives on Environmental Impact Assessment*, Rheidel, Dordrecht, pp. 213-231.

Wathern, P. (1988), Environmental impact assessment : preventive policy in practice, In : U.E. Simonis (ed), *Praventive Umweltpolitik*, Campus, Frankfurt am Main, pp. 167-185.

Wathern, P., S.N. Young, I.W. Brown and D.A. Robert (1986), Ecological evaluation techniques, *Landscape Planning* *12*, 403-420.

3

Ecological Crisis and Drought in Orissa

*Dr. Aruna Pillai**

Introduction : Ecological Crisis and Deforestation

It is necessary to understand the link between environmental degradation, deforestation and drought. Studies indicate that drought is not merely a natural calamity but due to man-made factors. The severe drought that affected the country in 1987-88 caused considerable loss of life and property. It was a combination of meteorological (shortfall in rain of over 20%), hydrological (depletion of groundwater) and agricultural drought (crop failure of more than 20%). In certain parts of India like tribal areas of Orissa, it was the fourth or fifth year of drought. The frequent occurrence of drought and floods are attributed to indiscriminate felling of trees.

Why does deforestation lead to environment deterioration ? Because it is the forest that maintains the ecological

*Senior Fellow, Centre for Research on New International Economic Order, Madras

balance needed for rain, conserving water in the ground and maintaining soil fertility. Trees protect the soil from erosion and keep up the water level. They release oxygen, consume carbon dioxide and transmit moisture to the atmosphere. Changes in vegetative cover can increase or decrease the run-off from a watershed. It can also alter the amount of rainfall in an area. Ecological balance is destroyed through deforestation and manifests itself in many so-called natural disasters.

Deforestation is one indication of the overexploitation of natural resources. The National Remote Sensing Agency released some figures based on the interpretation of satellite data. It shows that in 1972-75, 16.89% of the total land area of the country was under forest cover. By 1980-82 this had fallen to 14.10%, indicating an average loss of 1.3 mlllion hectares of forests every year. The optimum forest area required for ecological balance is placed at 33% of total land area; India's forest area is 22% of which not more than 10% is under adequate tree cover.

The process of deforestation started under British rule. But development policies followed in the last forty years (with the accent on heavy industries) have also contributed to massive deforestation. Most raw materials for industry come from natural resources. In addition to meeting the industrial and commercial demand for raw materials, forests provide revenue to the government. All these requirements have led to increasing denudation of forests and attempts at generation have not been able to keep pace. In Orissa for instance, 40% of the land area is supposed to be under forests; it is even higher in districts like Kalahandi, Koraput, Phulbani, Ganjam and Dhenkanal. But recent satellite image studies have shown that only 20% is under tree cover.

The majority of the tribals and forest dwellers in these areas live in conditions of poverty and destitution. The scheduled tribes who form 22.4% of the total population of the state are found mainly in Koraput, Kalahandi, Bolangir, Sambalpur, Phulbani, Ganjam, Mayurbhanj, Keonjhar and

Dhenkanal. The scheduled tribes together with the scheduled castes form 37% of the total population. They are socially and economically backward. These people live in villages in hilly areas and depend on shifting cultivation and minor forest produce from the forests for their survival.

Deforestation has had an adverse impact on tribal economy and culture. Studies in Orissa show that three decades ago, when the forests were plentiful, they provided the basic requirements of the forest dwellers such as food, thatching material, herbal medicines, fodder, fuel, etc. But unchecked clearfelling due to the needs of industry and commerce and introduction of the contractor system has led to depletion of forest resources in areas like Kalahandi and Phulbani. The increase in the reserve forest area has also resulted in the reduction of availability of forest produce for the tribals. Moreover, the ecological changes resulting from deforestation have led to increased frequency of drought. Earlier, they could depend on fruits, roots and tubers as well as meat of wild animals. Today these forest foods have almost disappeared leading to hunger and starvation among the tribals, particularly in the lean season.

Formerly, shifting cultivation had an 18-20 year cycle which has now been reduced to 3-6 years due to decrease in land available for cultivation. Within this shorter cycle, the tribals are unable to leave the land fallow for a period long enough for regeneration. Consequently, this form of cultivation leads to deforestation. Apart from food and land, deforestation by industrial clearfelling also deprives tribals of raw material required for crafts like basket weaving. As a result, they are giving up their traditional occupations and taking to selling firewood to urban areas through contractors. The increased dependence of the tribals on outsiders over a period of time, has led to indebtedness, land alienation and bondage to moneylenders and middlemen. In order to improve the situation of the tribals, who are struggling for survival, solutions have to be found for the basic problems of exploitation and impoverishment.

Drought Relief vs. Drought Action

Drought is defined as scarcity of foodgrains, water and fodder. The solutions sought by the government are based on the premise of "scarcity", thus dealing with the symptoms instead of the causative factors. The official agencies as well as the NGOs provide short term relief in the form of free meals, temporary employment, digging wells and installing pumpsets. Such schemes tend to cater to vested interests or they are populist in nature.

Large parts of Western Orissa covering Koraput, Bolangir, Sambalpur, Kalahandi, and Phulbani have been in the grip of severe drought for three consecutive years. Marginal agriculture practised on small uneconomical holdings is characteristic of this region. The cultivation of rice in the kharif season is almost entirely dependent on rainfall. Rice is the preferred cereal and the kharif harvest must be stretched till the next harvest. A good part of the harvest is kept for home consumption and only a small part is sold. During the rabi season, minor millets and oil seeds are grown. Both the harvests have been badly hit in last year's drought. Most cultivators are short of cash. They have no reserves or cushioning capacity. Such a fragile and unstable economy is not able to withstand the impact of drought.

Drought has led to a general lowering of standards of health and nutrition. There is the incidence of hunger that affects the poorer sections like scheduled castes and tribals, particularly the landless, and weaker groups like the aged, and women and children. Drought has also led to rising indebtedness and inability to repay loans. There has been a great deal of sale of assets—cattle, goats and personal assets like household utensils and silver jewellery. People take recourse to traditional coping mechanisms like migration, which has sharply increased in recent times. There are two types of migration. There is the short duration migration when people migrate from the drought hit areas of Bolangir to the irrigated areas of the district in the east. There is also the long term migration to cities such as Hyderabad, Madras, Lucknow, Dehra Dun etc. Agents come to the village and recruit labour against cash advances. This type of migration which is linked to scarcity and drought is taking place extensively in Western Orissa.

People try to engage in whatever non-agricultural work is possible. They collect and sell kendu leaf from the forest or make baskets and brooms from jungle grass which they sell at a pittance. In normal times, people used to depend on forests for their living as it provided them with fruits, tubers and other food. But deforestation has led to a destructive use of forest wealth. Now people are left with no other way of earning a living except to chop and sell fire-wood from the forests.

Government relief measures have been largely in the area of public feeding and employment. A massive feeding programme in Kalahandi has acted as a critical food supplement. Drought relief employment scheme however, do not generate the level of employment that could meet the demand for work. A scheme of well digging called the Jeevandhara has met with some success. This provides immediate employment and long-term benefits. It is mainly for small farmers from the SC/ST category. In Koraput and Kalahandi, a new scheme has been launched—ADAPT (Area Development Approach for Poverty Termination). It is basically a development plan with the following components : (1) a feeding programme (2) soil conservation (3) afforestation (4) road communication (5) lift irrigation points and minor irrigation works (6) health coverage. It is a coordinated effort to tackle the problems of rural poverty and the drought crisis. It is still in its initial stages of implementation.

In the voluntary sector, there are many NGOs engaged in drought relief work in Orissa. Some of them have realized that purely relief oriented measures or ad hoc operations are not enough to deal with the root causes of drought. The ecological crisis (of which drought is a manifestation) is a complex phenomenon which requires a long-term perspective and action on many fronts. Drought has had a destructive effect on the economic and socio-cultural environment of the rural poor. These people (mainly small farmers and landless labour lelonging to SCs/STs) living below the poverty line have no alternative means of livelihood and are unable to bear the brunt of the damages caused by drought. They also do not possess the socio-political bargaining power to change the situation. Any meaningful drought action should be oriented towards the struggle of the poor to improve the conditions in which they live and to build up their resistance against the degradation of the natural resources like forests, grazing lands, village tanks and other common property assets. The destruction of this natural environment which is the very base of their survival is one of the primary causes of drought.

Our Centre (the Centre for Research on New International Economic Order) which is engaged in development research, has brought together a group of NGOs working in some of the drought-affected villages of seven districts in Orissa : Koraput, Kalahandi, Bolangir, Phulbani, Ganjam, Dhenkanal and Mayurbhanj. We have evo'ved a set of criteria for long-term drought action programmes. This alternative drought action is based on people's participation and aims at building up the organisation of the poor at the village level. The drought action is directed towards the most marginalised groups, among them being small and marginal farmers and landless belonging to SCs/STs, with special focus on women and children. Drought action is to be an integrated effort to meet the basic needs of the poor, with high priority to ecological measures-e.g. afforestation, dry-farming, soil and water conservation etc. Drought action includes measures to counteract land alienation, distress sale of cattle, mounting indebtedness. Steps will be taken to strengthen people's own ability to take appropriate steps to face drought. Drought action will contribute to social change and social justice in favour of the poor and marginalised. Based on these criteria, NGOs in the Orissa drought action forum have already begun to implement their action programme.

Participatory Action Research

Self-reliance and people's participation form the base of the alternative drought action programmes being undertaken by the NGOs. However, there is much exploitation and oppression of the vulnerable sections of the rural population, such as the landless, tribals and women. This being the case, self-reliant development is not possible unless the poor understand the socio-economic reality of their environment, especially, the forces responsible for their poverty and oppression. Participatory action research is a method of conscientization and mobilization. It is responsive to the needs of the exploited and is committed to helping them to understand their situation and change it. It is a process of awareness building, social transformation and action.

Our Centre has carried out a participatory action research programme in Orissa in the seven districts where the NGOs who are members of the drought action forum are working, The drought research field staff are full-time workers of their respective NGOs. They are not doing research as a separate activity. It is carried out along with their development work and action programmes. Research and action are very much inter-linked. The field work has covered the following areas of research :

- root causes of drought and coping mechanisms of past
- impact of land alienation processes and modern development strategies on the marginalised
- socio-economic effects of drought, impact on women and children and responses to drought by government and NGOs
- village and household survey

In the drought research programme, the emphasis has been on knowledge generated and returned to the people. For instance, information is gathered through informal discussions with the people, both in groups and individually over a period of time. This has yielded a variety of findings and insights.

During the initial three months, the field staff studied the root causes of drought and coping mechanisms of the past. It was found that according to the people, most of the villages were surrounded by dense forests about thirty years ago. Due to population pressure and urban encroachment, people had to cut the forests in order to acquire cultivable land.

People had begun to perceive that depletion of forest cover caused hardship to human and animal life, eventually led to reduced rainfall and to a hotter and drier climate.

Some perceived drought and hardship as a curse of God, and the result of their failure to appease the village deity with the adequate offerings.

The older generation felt that in the past, when the barter system was more common, and each family could obtain their requirements by trading or exchanging their produce or services, the village community was self-sufficient.

The cash economy along with the entry of traders and money lenders from outside has adversely changed people's way of life. People said it would not be possible to restore the old system now. Moreover, many of the younger people prefer the cash economy and would like to travel to towns and to be integrated into the mainstream.

Information was collected and analysed by the field staff and the findings returned to the people through village assemblies, group discussions, songs, skits, dances and dramas, depending upon the communication skills of the field staff. During these awareness building exercises, the field staff and the people engaged in a dialogue and shared their experiences relating to drought. The action programmes of the NGOs in their area were discussed and people expressed their preferences for the projects and programmes they required.

The aim of the participatory action research programme is to involve the people in the research, identify their needs and help them find solutions to their problems.

Need for Sustainable Development

After forty years of the development strategy based on industrialization and growth, the gap between developed and developing countries has widened. The view is no longer held that developing countries can attain the standards of living obtaining in the West. The failure of the development paradigm has led to a rethinking of its goals and a shift in emphasis from growth to redistribution with growth. Development is no longer viewed in purely economic terms. Development should also lead to a decline in unemployment, inequality and poverty as well as an increase in per capita income.

The pursuit of limitless growth has been questioned. Development has depended upon the harnessing of science technology to achieve the projected increases in GNP. Most of these technologies involve the consumption of energy from non-renewable sources. These natural resources are being depleted at a rapid rate, without considering the consequences of environmental degradation and ecological imbalance. Indiscriminate use of fertilizers, effluents from monern industries, destruction of forests, intensive mining, air and water pollution —all of which cause environmental problems have been brought on by pursuing a policy of growth at all costs.

So far, the development process has shown scant regard for the environment. Development has been concerned only with the increasing demand for natural resources for the production of goods and services. This resource-intensive path has led to serious disruption of ecological balance. An element that is missing from most planning and development is sustainability. Most developing countries are trying to follow the Western mode of development without any idea of the ecological costs involved. Environmental awareness has to be developed in Third World countries in order to make their development sustainable and as self-reliant as possible.

The concept of outer limits is related to the condition of sustainability. Outer limits refer to resources—both non-renewable and renewable—and environmental systems. Non-renewable resources are those which are not replaceable by nature or man. Renewable resources are those, which under certain circumstances will replace themselves or can be replaced by man. Environmental systems are more difficult to define. There are a number of ways outer limits could be reached in environmental systems. Man plays a role in determining where outer limits are with relation to population, food, energy and other essential resources. The responsibility for the proper utilization and management of natural resources and environmental systems rests with man. The challenge lies in meeting basic human needs without transgressing outer limits.

The new awareness of environmental problems has led to the realization that environment is a dimension of development. Problems of resources, energy, environment, population and development cannot be viewed in isolation; they should be examined in relation to each other. This, in turn, has led to the concept of eco-development, which can be broadly defined as development on ecologically sustainable principles. Emphasis should be laid on satisfying basic needs relating to food, housing, health and education. Efforts should be made to develop those resources which are available within the eco-region. In the interests of equity and social justice, due regard should be paid to the equitable utilization of natural resources. These resources should also be utilized in a rational manner with a view to protecting the interests of future generations. It is the absence of these measures of conservation that lead to ecological crisis.

References

Agarwal, Anil, Ravi Chopra and Kalpana Sharma, The State of India's Environment : A Citizen's Report, Centre for Science and Environment, New Delhi, 1982.

Agarwal, Anil and Sunita Narain, The State of India's Environment : A Second Citizen's Report 1984-85, Centre for Science and Environment, New Delhi, 1985.

Agarwal, Anil, "Development Policies, Ecology and Gandhi's Last Person", Social Action, 37 (No. 1, January—March 1987), pp. 45-60.

Bandyopadhyay, J., "Political Ecology of Drought and Water Scarcity, Need for an Ecological Water Resources Policy", Economic and Political Weekly, December 12, 1987, pp. 2159-2169.

Bandyopadhyay, J., and Vandana Shiva, "Political Economy of Ecology Movements", Economic and Political Weekly, June 11, 1988, pp. 1223-1232.

Bandyopadhyay, J., "Water Scarcity by Choice", Seminar 346 (Drought) June 1988, pp. 24-29.

Bhatia B.M., Indian Agriculture : A Policy Perspective, Sage Publications, New Delhi, 1988.

Dhanagare, D.N., "Action Groups and Social Transformation", Lokoyan Bulletin 6 : 5, 1988, pp. 37-59.

Dhawan, B.D., "Management of Ground Water Resources, Direct verses Indirect Regulatory Mechanisms", Economic and Political Weekly, September 5-12, 1987, pp. 1553-1554.

Dhawan, B.D., "Indian Irrigation : An Assessment", Economic and Political Weekly, May 7, 1988, pp. 965-971.

Dhawan, B.D., Irrigation in India's Agricultural Development, Sage Publications, New Delhi, 1988

Dube, S.C., Modernization and Development, The Search for Alternative Paradigms, Vistaar Publications, New Delhi, 1988.

Fernandes, Walter and Rajesh Tandon (eds.) Participatory Research and Evaluation : Experiments in Research as a Process of Liberation, Indian Social Institute, New Delhi, 1981.

Fernandes, Walter (ed.), Inequality, its Bases and Search for Solutions, Dr. Alfred de Souza Memorial Essays, Indian Social Institute, New Delhi, 1986.

Fernandes, Walter, "Drought in India : Its Causes and Victims", Social Action, 37 (No. 4, October—December 1987), pp, 421-443.

Fernandes, Walter, "The Forest Policy and the Forest Conservation (Amendment) Act, 1988", Lokayan Bulletin 7 : 1, 1989, pp. 37-46.

Fernandes, Walter, Geeta Menon and Philip Viegas, Forests, Environment and Tribal Economy : Deforestation, Impoverishment and Marginalisation in Orissa, Indian Social Institute, New Delhi, 1987.

Fernandes, Walter, "Participatory Research and Action in India Today", Social Action 39 (No. 1, January—March 1989), pp. 1-21.

Freire, Paulo, Pedagogy of the Oppressed, Penguin Books, London 1972. Reprint 1980.

Haque, Wahidul, Niranjan Mehta, Anisur Rahman and Ponna Wignaraja, "Towards a Theory of Rural Development", Development Dialogue, 1972 : 2.

Jain, L.C., "Poverty, Environment, Development, A View from Gandhi's Window", Economic and Political Weekly, February 13, 1988, pp. 311-320.

Matthews, William, H. (ed.), Outer Limits and Human Needs, Resource and Environmental Issues of Development Strategies, Dag Hammarskjold Foundation, Uppsala, 1976.

Meher-Homji, V.M., "Deforestation, Ecological Destabilisation and Drought", Social Action, 38 (No. 2, April—June 1988), pp. 115-128

Mooney, Pat Roy, "The Law of the Seed, Another Development and Plant Genetic Resources", Development Dialogue, 1983 : 1-2.

Murishwar, Joy and Walter Fernandes, "Marginalisation, Coping Mechanisms and Long-term Solutions to Drought", Social Action, 38 (No. 2, April—June 1988), pp. 162-178.

Nadkarni, M.V., "Agricultural Development and Ecology—An Economists View", Indian Journal of Agricultural Economics, July—September 1987, pp. 359-375.

Nerfin, Marc (ed.), Another Development : Approaches aad Strategies, The Dag Hammarskjold Foundation, Uppsala, 1977.

Paul, Samuel, "Poverty Alleviation and Participation - The Case for Government—Grassroots Agency Collaboration', Economic and Political Weekly, January 14, 1989, pp. 100-106.

Qureshi, Moeen, A., "The World Bank and NGOs : New Approaches", Lokayan Bulletin 6 : 4, 1988, pp. 45-54.

Rangasami, Amrita, "The Limits of Drought", Seminar 346 (Drought) June 1988, pp. 30-35

Rao, Hanumantha, C.H., "Agricultural Development and Ecological Degradation ; An Analytical Framework", Economic and Political Weekly, December 24-31, 1988, pp. A-142-A-146.

Wignaraja, Ponna and Akmal Hussein (eds.), The Challenge in South Asia, Development, Democracy and Regional Co operation. Sage Publications, New Delhi, 1989.

4

Pollution Taxes—A Way Cut for Environmental Protection

*A.C. Angrish**

In the Paper, an attempt has been made to highlight the economic significance of the taxes on pollution. In Part-I, the purpose of such taxes has been specified. In Part-II, the economic justification has been discussed. Part-III, clarifies the academic discussion on pollution taxes along with a comparison of unilateral tax approach and the bilateral tax approach. Part-IV concludes the discussion.

Taxes on pollution have assumed importance in view of the growing environmental degration and the widely felt need to provide environmental protection. "The Sulfur and Nitrogen Emissions Tax Bill of 1987" in the United States of America, introduced by Congressman Judd Gregg has developed a new interest in taxes on pollution since they can also be good revenue raisers.

In principle, taxes on pollution serve two purposes : (i) as

* Professor, University of Jodhpur, Rajasthan.

effective policy approach to protection of the environment and (ii) as a good revenue source to the Government. The taxes on pollution help in increasing social welfare and cost society less than the benefits they provide.

Environment is a scarce resource. Its limited capacity to absorb waste and to avoid congestion is a proof of this fact. Earlier "air" was classified as a free good i.e. it was considered to be unlimited in supply. In effect, this led to an absence of an appropriate price for it and subsequently to its overuse. The result has been "Air Pollution". In a situation of pollution, the free market system may not reflect true social cost.

The taxes on pollution serve to correct a serious source of market failure; the absence of a "Price" needed to prevent the careless and excessive use of scarce environmental resources (Oates, 1988).

A factory which emits smoke on a neighbouring residential area imposes real costs on those who have to live with the dirty air. But the factory may not bear this cost in the way it bears while employing resources like labour and raw materials. The price of latter resources shows the way to economise on their use, but there is no incentive for the factory to control its smoke emissions in the absence of an appropriate price to restrict its use to excess. The direct policy implication that follows is that there is a need for governmental intervention and in the absence of free market system to operate, the government should impose a tax on damaging waste emissions.

The success of any environment programme shall depend on its capacity to discover and employ more effective and less costly methods of pollution control. The pollution taxes provide incentives in encouraging the selection of proper control technology and for developing effective abatement technology. Such methods could go a long way in providing direct benefits to polluters by reducing their costs, and may also benefit the society at large through cleaner environment that would follow.

Pollution taxes have given rise to a lot of academic discussion by prominent economic thinkers. They have argued their view-points with reference to unilateral tax approach and bilateral tax approach. Should the polluters alone be taxed ? Or should be victims of pollution be taxed also ? are the important questions dealt with by them.

Professor Pigou (1932) while introducing the concept of pollution tax, has taken the example of a rail road producing service and emitting an amount of engine sparks which cause fire and property damage or increase the production cost of raising crops on the nearby farm land. His approach specified that the socially optimal level of services would be provided automatically by the rail road if the government imposed a unilateral pollution tax. The rate of tax may be set equal to the value of damages that the marginal unit of engine sparks generated at the socially optimal output level of the service.

Professor Coase (1960) on the other hand, has stated a bilateral tax system to control pollution. He argued that the behaviour of not only the polluters but also the victims of the pollution should be restricted. He argued that if the owner of a factory emitting smoke made to pay a tax equal to the damage caused, the nearby residents should also be made to pay a tax in an amount equal to the additional cost incurred by the owner in order to avoid the damages. He pointed out that because the victims of pollution decide to reside close to the factory, the latter is bound down to incur extra costs.

While commenting upon the undesirability of the pigouvian tax system and relying more on negotiation, Professor Coase reasoned out that the negotiations between the rail road and the farmers can solve the pollution problem. The rail road may offer compensation to the farmers for the damages it caused by emitting engine spark and the farmers may accept it if the compensation was sufficiently large to cover the profits that they would lose by granting permission. Alternatively, the farmers may offer the compensation to the rail road for reducing its sparks as long as the amount of compensation was smaller than the amount of profits that could be secured from selling the crops saved from fire throgh the reduction in engine sparks.

The Coasian analysis is thus based on voluntary negotiations between the parties so long as the room for negotiation exists. But if adverse reactions take place resort to bilateral tax system may become desirable. The Pigouvian approach, however, requires the presence of the Government to administer the tax. In the coasian system, the polluters or the victims, whosoever receives compensation has no incentive to minimise their damages because the party on whom the liability is placed is fully compensated for the damages, while under the pigouvian tax system, the victims have every incentive to minimise their damages, for they could gain nothing by increasing damages.

Professor Baumol (1972, 1974) argued that a tax imposed on the factory owner would determine the amount of smoke emitted and the amount of smoke thus determined would in turn control the number of people who established residences in the neighbourhood of the factory, for they would take the smoke into account in choosing their locations. Baumol accordingly contended that all that is needed is a wisely chosen efluent tax imposed on the generators of pollution alone will achieve a socially efficient solution and that the victims of pollution, if left uncontrolled, will arrive at optimal solution in response to the prevailing pollution level determined by the tax.

His analysis is no doubt sound within its theoretical framework but it seems rather difficult to implement it in reality unless the government assumes an important role in swinging the outcome to its desired level.

The government may adopt a number of alternative as corrective measures, with regard to pollution control. Laying down certain "standards", setting down regulations, making it obligatory to hold licences etc., could be some of the measures. It may subsidise investments in pollution-control technology. It may impose fees, fines or penalties for the discharge of various pollutants. Itmay also restrict the use of certain fuels that generate excessive pollutants or may put a heavy pollution tax on them.

Although the aforesaid alternatives may be useful in reducing the level of pollution but the distributional implications of such policies could be quite different. From that angle, out of various forms of control, pollution taxes appear to be preferable policy tool for environmental protection.

If the government imposes pollution taxes, the polluters shall directly bear the costs of pollution abatement while the users of the commodities (produced by the polluters) shall indirectly bear it. The tax revenue could be used to compensate those who are damaged by the pollution. It shall contribute to increase in the economic welfare.

To make them effective, it is necessary that the pollution taxes be levied on the polluting activity itself. For example, taxation on the sulfur emissions directly is needed rather than taxing the coal with a high sulfur content. The taxes could be adjusted to reflect changing costs, geographical variations etc., so that they may become more flexible in nature in protecting the environment.

References

Baumol, William J., "On Taxation and the Control of Externalities", The American Economic Review, LXII, No. 3 (June 1, 1972) : pp. 307-322.

Baumol, William J., "On Taxation and the Control of Externalities : Reply", The American Economic Review, LXIV, No. 3 (June 1974) : p. 472

Coase, Ronald, "The Problem of Social Cost", Journal of Law and Economics, III (October, 1960) : pp. 1-44.

Oates, Wallace E., "Should Pollution Be Taxed," Economic Impact, No. 4, 1988 : pp. 27-31.

Pigou, A.C., The Economics of Welfare, 4th Edition, 1932.

5

Integrating Environmental Factors into the Land-use Planning Process

*Maria do Rosario Partidario**

Stating the Scope

The importance of considering an overall perspective of the environmental system in the land-use planning process is becoming widely recognized. As Lichfield and Marinov (1977) once wrote, environmental concerns in land-use planning "cover not simply the traditional regard for amenity—it extends to the social and economic conditions in which man lives in settlements, and in this sense the basic economy of a town can also be fundamental to its environmental concerns". Several papers and reports have been produced and research is being carried out in several parts of the world to look at the most feasible and effective way of bringing together, in an integrated way, the environment and the land-use planning.

In a research project being developed at the University

* Professor, University of Aberdeen, U.K.

of Aberdeen on the role of Environmental Impact Assessment (EIA) as an assessment tool in the land-use planning process, it is felt that one of the basic steps should be to integrate environmental factors in the land-use planning process before the enforcement of any EIA framework. This is not an easy task, though, and a number of difficulties may arise when attempting this integration. This paper focuses on some practical conditions that should be met to achieve this implementation.

In any political or planning system, there is always a transition problem between a theoretical or scientific approach to a topic and the effective development of its implementation. A well designed and adaptative procedure does not necessarily produce the best results if the framework is not robust.

On the other hand, if existing conditions to implement environment or planning approaches are not clearly identified, then all the efforts made on the development of theory and scientific research may well be a waste of time. This does not mean that the most perfect conditions for implementation must be established, but it must, at least, provide the grounds on which any technical proposal put forward will be made more feasible.

Thus, some problems and issues that should be tackled when integrating environmental factors in the land-use planning process will be presented for discussion. From evidence available, these are the most usually referred conditions to the effectiveness of any possible approach.

A broad definition of the environment will be presented, before some institutional, political and technical conditions will be listed that should make more feasible and productive the integration of environment concerns into the land-use planning process.

A Broad Concept of the Environment

The word environment has become very fashionable and

commonly used over the past twenty years. This is a result of the environmental movement and the subsequent growing interest in environmental issues such as conservation and protection, pollution control, environmental quality and assessment. Its importance in the technical, political and common vocabulary has led to a diversity of ways in which it can be used and to a certain confusion and subjectivity concerning its real meaning.

There are those who still consider the environment as the biological or ecological sector, misunderstanding the meaning of the words environment and nature. This represents a very narrow and restrictive meaning.

The environment is also quite often used when referring to the physical disturbances caused by antropogenic activity—environmental pollution. The scope, in this case, includes four main components : air pollution, water pollution, noise pollution and soil or land pollution. Although not so restrictive as the previous definition, the environment in this case only refers to the physical expression of the effects between men's activities on the surroundings. It is much more comprehensive but there are still many other components that can be identified within the environmental system.

A broader definition is one that understands the environment as a comprehensive and dynamic system, including a complex set of different but interdependent elements. These elements will be aggregated in components broadly identified in physical, ecological, social, economic, institutional and political components. It is based on a complex network that represents the relationships between the different layers of the several components within the environmental system hierarchy. Each of these components, or sub-systems, are characterized by particular issues and act as unbounded systems. Any modification in one of the sub-systems will have effects in all the other sub-systems, depending on the type of the established relationships. Eventually, all the environmental system will be reacting in some way. Figure 1 shows a two dimensions representation of the sub-systems integration within the environmental system.

Although highly complex, this broad approach to the environment is adopted in this paper. The significance of each reaction in relation to the effect on the whole environment will be different. To avoid the complexity of the system and work on more feasible grounds, it will be necessary to focus on the most significant features that influence the environmental behaviour by changing its dynamics.

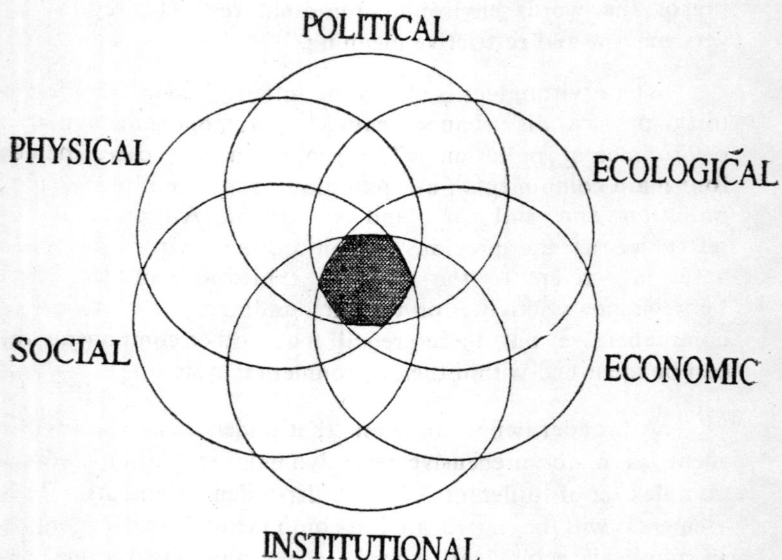

Fig. 1. Main interdependent components of the Environmental System

Conditions for Integration

It is time to look now into several conditions that, if not met to a greater or lesser extent, may become severe constraints on the integration of environmental factors in the land-use planning process.

1. Existence of Systematic and Available Data

Quite often the lack of appropriate baseline studies is referred as a severe limitation on adopting a comprehensive and efficient approach. However, the production of massive volumes of non-organized, scattered data, does not help. It demands a lot of technical staff time and effort and represents a waste of human and financial resources. The development therefore of adequate environmental databases will create and facilitate the manipulation and analysis of systematic data and subsequently make it available to provide relevant and useful information.

According to Pocock (1981), environmental baseline studies in plans are important in terms of the :

— "selection of appropriate indicators;

— availability of relevant information;

— presentation of the information in a readily, comprehensive, concise, meaningful and accurate way, allowing comparisons between areas to be made;

— inferring of trends;

— inclusion of knowledge on causal factors."

The data thus obtained will enable the description of the existing environment by producing information of soils, topography, vegetation, transport, human settlements, land use, scenic and historical areas, air, water and noise pollution, solid waste, natural hazards. The collection, mapping and analysis of this information will then according to Walker (1977) enable :

—" the determination of land suitability for development;

— the identification and analysis of cause and effect relationships to development actions;

— preparation of site selection, design criteria and performance standards for future development types, e.g. residential, open space, transportation uses;

— monitoring environmental changes and evaluate development on a continuous basis."

Reference must be made to the establisment of monitoring programmes which are necessary to keep a continuous record of environmental change and to enable time-series environmental baseline studies to be provided.

2. *Development of Geographical Information Systems (GIS)*

One of the most important applications of GIS lies in its capability to store, manipulate, analyse and produce data and/or information. Computer-based GIS have even more advantages for all these capabilities are improved owing to the enormous reduction in time consumption.

One of the problems of non-organized information is the "inappropriate use of the data (or information) when rather ambiguous, leading many times to misinterpretation to suit particular prejudice" (O'Rioordan 1979).

GIS not only provides the setting for an environmental database as it facilitates its use and still has the potential for acquiring new and unique information by querying the database. Information organized in this way can be more easily updated and, at the same time, provide systematic reporting of on-going development changes.

GIS can also be organized to focus on critical limits of key resources. It will also enable the identification of operational thresholds where further major investment must be made by the community to avoid environmental damage or to sustain further development (Hall, 1976). The development of a set of environmental indicators within this framework, coupled to different policy guidelines will provide a standard of measurement for evaluation purposes.

3. *Definition of Environmental Criteria*

Criteria could be defined as principles or standards on the basis of which something is judged. As a standard it can have a similar meaning to thresholds, which define physical limitations to development (Kozlowski and Hughes, 1972). As a principle, it represents a set of rules based on value judgements, reasoning and common sense, therefore becoming the expression of community wishes.

According to Pinho (1985) two basic factors are involved in the setting of environmental standards : community goals and scientific criteria. If the environment is to be integrated into the land-use planning process, environmental standards must be translated into environmental criteria as an aid to more rational decision-making. For Albano (1990) an integrated approach will be based on "criteria that will confer more transparency to planning decisions and, at the same time, will consider all aspects involved in planning".

The set of environmental criteria to be developed must depend, however, on the level in the tier of plans to which it is applied and, mainly, on the community values and environmental characteristics of the area. It must then be developed and adapted to each particular situation.

As an example, let us consider the case of a local plan for an urban development. The environmental criteria likely to be developed could include :

— density and housing type

— use of local building materials

— quality of water streams for urban recreation

— availability of urban green areas for recreation

— efficiency of the sewage treatment system

— accessibility to social facilities

— traffic allowed in certain areas of town (in terms of category and number of vehicles) to ensure security, noise and air quality levels.

4. *Making Use of Expert Systems*

The development of expert systems may have considerable advantages to the decision-making process, since it will enable environmental information to be used in an appropriate and optimal way.

Expert systems are computer-based frameworks that "use specialist knowledge consisting of facts and rules of thumb (heusistics) which are combined to produce new knowledge" (Leary, 1988). Leary (1988) defines expert systems as "programmes which manipulate knowledge rather than numbers".

The most important objective is to use expertise that can provide a more flexible and adaptative way of making decisions. The idea of knowledge consisting of a set of rules may seem rather inflexible especially in some areas of planning such as urban design and development control. But if one thinks on the knowledge base as containing the wisdom of experts who have transformed, through experience, an all set of complex facts and strategies into more manageable simplified forms, then they will be able to assess a problem and come up with an adequate solution quickly (Leary, 1988). Considering the diversity of situations and the wide range of environmental attributes that one deals with when developing land-use plans, the existence of a panel of experts, or computer-based expert systems will provide the grounds for a more rational argument for development decisions.

5. *The Political Commitment*

Active and direct political involvement is especially important since most "environmental quality conflicts are much less technically or economically determined then they are politically determined" (Pinho, 1985).

Research and development concerning the environment requires political commitment. The provision of government funding for the development of research projects is one form of political involvement. On the other hand the definition of environmental goals and objectives is a political decision as much as the development of environmental control policy instruments. These represent key stages in the decision-making process that will orientate further actions. Their contribution is therefore determinant is establishing an anticipatory and prospective approach towards development control. It will enable adequate spatial coordination of potential polluting and sensitive land uses, which at the same time consider natural, social and economic issues.

Political concern on environmental issues should also be expressed in final decisions. Planners do not make final decisions but have to ensure that they are made by elected representatives who have environmental awareness directed towards more comprehensive planning.

6. Institutional Fragmentation

Another major factor that is linked to political commitment is fragmentation among institutional agencies. This fragmentation can be seen in three different facets. First, from evidence available, in many countries there is still a lack of linkage between the planning and the environmental affairs departments, especially at the national level. Secondly, it is concerned with the lack of linkage between the three main levels of action : the national, regional and local level. And finally, it reflects the sectoral organization of the administration, where the planning and environmental control responsibility is divided among specialized and independent agencies established to deal with particular problems.

"The sectoral approach by a number of non-coordinated government bodies reflects an ill-coordinated management and determines a set of conflicting sectoral policies" (O'Riordan, 1979). The absence of effective inter-governmental or inter-agency coordination leads to a "narrowly focussed, single-purpose functional planning" (HUD, 1977). One of the problems of this sectoral approach is that measures designed to solve a particular problem in one sector may often result in the creation of cumulative effects in other planning sectors.

Efforts should be made for more direct cooperation between the environmental and planning authorities and better coordination between national, regional and local levels and amongst all the sectoral government agencies involved.

Attempts to coordinate the various planning programmes having impacts on the environment should be developed. Programme managers should be encouraged to recognize the work performed in one programme to fulfil the requirements of another (HUD, 1977). Coordination will enable joint action to develop more consistent legislation, common environmental criteria and methodological or procedural definitions and recognition of mutual restrictions and constraints (DGQA, 1988).

7. *Improvement of the Planning System Flexibility*

Planning systems include :

— the planning and decision-making process

— the relationships between agencies

— policy and land use control instruments

— public participation procedures

— monitoring of plan implementation.

The flexibility of the planning system is commonly referred as a key requirement within the process of integration of environmental factors in the land-use planning.

Lichfield and Marinov (1977) stress the need for flexibility in land-use planning and control to enable the process to reverse. They argue that predictions do not allow sufficient confidence to assert that all developments in the future should conform to the plan. Therefore, only through a flexible planning system is it possible to work under "conditions of uncertainty" and yet adopt an adaptative approach, given environmental and planning constraints.

Plans that are out of date and technically inadequate are the main reasons presented by Booth (1984) to justify "the inflexibility of plans, despite review". To meet changes more adequately, Booth emphasis the need for increased flexibility through the development of less structured methodologies an impact analysis. Increasing flexibility will provide and informal and more dynamic process of decision-making (Pinho, 1985), within which more flexible methods of assessment should be carried out (Albano, 1990).

8. Environmental Awareness

The last but not the least, environmental awareness provides the grounds for a participatory environmental comprehensive planning.

Consultation and public participation is one of the most important stages within both the plan-making and the EIA process. Nevertheless, it is often considered to be a difficult and conflicting process. When the general public and the statutory or non-statutory agencies are called to give their opinion on planning proposals or decisions, they often reflect self interests such as the protection of private interests and property. This attitude is often a result of a lack of understanding and knowledge of environmental problems, issues and aims.

Environmental training should not only be directed at public, statutory and non-statutory agencies but also extended to the political sector to provide the basis for increased political commitment.

Conclusions

The traditional land-use decision-making process tended to be dominated by economic criteria, resulting in environmental degradation, loss of amenity and ecological diversity and other unanticipated social costs. It failed to recognize that the environment comprises a complex set of interdependent natural and man-made elements with ecological processes constituting natural determinants of land-use.

The inclusion of environmental aspects into land-use planning will gradually provide the grounds for an assessment procedure which pays proper attention to physical, ecological, social, economic, institutional and political aspects. If the set of economic criteria is expanded to include more comprehensive criteria, this should help determine more optimal land-use allocations.

The environmental aspects constitute an important structuring element in land-use planning. Restrictions and constraints creating difficulties in the integration of environmental factors in the land-use planning process should then be carefully considered and appropriate measures taken to overcome any problem.

References

Albano, Antonio (1990), "Economic and environmental costs, An integrated approach to planning policy", VIA, 13 : 76-85.

Booth, A.G., (1984), "Ecology and Planning : the Planning Component", in Roberts & Roberts (Ed.), Planning and Ecology, Chapman & Hall, London.

DGQA (Direccao-Geral DA Qualidate do Ambiente) (1988), Ligacao entre as questoes da Qualidade do Ambiente e do Ordenamento do Territorio, MPTA, Lisboa.

Lichfield, N., and U. Marinov, (1977), "Land-use planning and environmental protection : Convergence or divergence ?" Environment and Planning A, 9 : 985-1002.

Hall, R., (1977), "Meirs—a method for evaluating the environmental impacts of general plans", Water, Air & Soil Pollution, 7 : 251-260.

Hud (Housing and Urban Development US Dept) (1977), Integration of Environmental Considerations in the Comprehensive Planning and Management Process, U.S., Department of Housing and Urban Development, Washington D.C.

Kozlowski, J., and J.T., Hughes (1972), Threshold Analysis, The Architectural Press, London.

Leary, Michael (1988), "What are expert systems and how can they contribute to planning ?", Planning Outlook, 31 (1) : 7-12.

O'riordan, T., (1979), "Ecological studies and political decisions", Environment and Planning, 11 (7) : 805-813.

Pinho, P., (1985), Planning, and Control for Environmental Policy : A study in Portugal, Ph. D. Thesis, Strathclyde University, Glasgow.

Pocock, R.L., (1981), "Describing the state of the environment at the urban scale" in Joyce, F., (Ed.), Local Government and Environmental Planning and Control, Gower : 93-110.

6

The Indian Culture and Natural Environment : An Ecological Analysis

*K. Ravi**

Introduction

"Peace to the sky;
Peace to the atmosphere;
Peace to the Earth;
Peace to the Water;
Peace to the Plants;
Peace to the Trees;
Peace to all beings;"

This was the prayer once offered by the people of India and still this prayer inspires many souls. Perhaps, among all the cultures developed in the world so far in the history of human civilization, it is the Indian culture which shows an ecological orientation of evolution. As the cultural manifestations can have its inherent attributes in the mantle of religion,

*Professor, Centre for Environmental Planning and Technology, Ahmedabad.

it is the basic governing principles of the religion which guide the individual and the society and hence the total manifestations of culture and its different dimensions. The cultural evolution of Indian people clearly shows that the basic approach of its people was complete submission to nature and the suremacy of the nature and subordination of man was an accepted doctrine under which the life evolved/revolved. When other culture considered man as supreme and nature as a subordinate factor (for the sustenence of man), the Indian people/culture maintained nature and its entire system as supreme and man as a dependant factor. They have realised that all the living things existing in the nature are the members of the family of nature and each living being have its right to live and maintain its system deriving all the necessary elements from the nature without hurting the other.

It is interesting to note that the cultural synthesis of Indian society is a symbiotic union between the Aryan and Dravidian cultures. However, almost all the ecological characteristics of culture which now observed in our cultural arena was basically originated from the Dravidian approach of life and enriched by Aryan culture for which the archeological remains of Harappa-Mohanjadharo remains as an evidence.

Dravidian Culture and Ecosystem

The reverence of the people of Harappa-Mohanjadharo to sun, the embodiment of energy and the ultimate preserver of ecosystem, and other elements in the ecosystem like trees, animals, snakes and birds shows that the existing ecological manifestations of Indian culture and way of life is mainly attributed to Dravidian people (or originated from Dravidian people). And interestingly enough though there exists some sort of similiar pattern in the religious and cultural practices and manifestations among the Indian people the ecological reverence of the Dravidian society is much more prevalent and rigid than of the Aryan society. For instance though, the snake worship is observed in some parts of India, the intensity of the reverence of this practice which we now observe in some of the Dravidian society like Kerala is exactly like that of Harappa—Mohenjadharo people—Supporting the reverence of Dravidian people to ecosystem, Dr. B.C. Sinha observes: "Indus people regarded bulls as most sacred, while vedic Aryans had the greatest respect for the cow. But no animal is so important and so widely spread as the serpant which is either religiously worshipped or feared"[1]. Again attributing the ecological reverence of Dravidians he writes, "if the Dravidians were the architest of the Harappa culture in the 3rd or 4th millenium B.C., the earliest example of tree worship can be seen in their seals. A pipal tree is depicted in one of the seals of Mohenjadharo. A seal from Harappa is engraved with the likeness of a weeping tree with hanging branches"[2].

In Kerala and Tamilnadu the pipel tree is symbol of peace, prosperity and manifestation of God and is being worshiped and it is an integral and unavoidable element of the temple ecosystem. Considering the ecological similarities of cultural manifestations between the Harappa—Mohenjadharo and eco-system oriented culture and religious practices in Kerala, the analysis of the ecological linkages and relationships of cultural and religious practices of Kerala society will be a representative sample of the total Dravidian society. Hence, such an investigation will lead towards identifying the environmental and ecological reverence and approach of Indian People in general, and Dravidian people, in particular.

The Kerala Culture and Man Nature Relationship

Perphaps no region in India where the man-nature relationship is so symbiotic and intense like that of Kerala. It is quite interesting to study whether people became more conscious of the delicate interactions and relationships of nature because of the fullest manifestation of nature and its versatile biotic productivity or because the peole adopted a life style which became more congenial and coherent for the sustained maintenance of ecosystem. One wonders whether the life of the people of Kerala is so ecosystem oriented that every aspect of life and culture is perfectly interwoven with the nature that no human action affect negatively the ecosystem but it is always a supporting factor for the furtherence of nature and its other elements. This symbiotic man-nature relationship one can observe in all aspects of Kerala life.

'Tharavad'* and Ecosystem

Though Throughout India joint family system is a characteristic manifestation of the total Hindu Society, no other society in India can claim an ecosystem oriented joint family system which is unique and distinctive of Kerala society. It is unfortunate that this uniqueness of Kerala's societal life was so far not unravelled/explained in an ecological frame works.

The joint family system (Tharavad) in Kerala evolved through ages was not a mere mechanism of the relationship among individuals but encompassing the entire network of relationship between man and nature. "Tharavad", the pivotal element of joint family is firmly interwoven with the nature. The 'Tharavad' is not merely refers to a built structure or house but an integrated ecosystem. The 'Tharavad' system consists of a group of people, a chain of ecological elements like temple, ponds, the entire agro-ecosystem which comes under the authority of 'Tharavad' and a chain of cultural and religious practices like tree cult and animal cult. The life and culture revolve around these factors perfectly in symbiotic union with the nature. In other words, it is the 'Tharavad' which function as the basic balancing factor within and between man and nature.

Tharavad as the Cradle of the Ecology Oriented Culture

In fact, it is the religious and cultural environment through which an individual or society happen to evolve is the guiding factor in maintaining the balance of ecosystem. This fact is all the while manifesting in the history of social and ecological

*'Tharavad' means the main joint family and this is a symbolic representation of the martiarohal family system prevailing in Kerala.

evolution of Kerala and the 'Tharavad' is the centrally controlling element. An individual born in the 'Tharavad' is first introduced to the natural elements like Sun, Moon, Rain, Trees, Animals, Birds, reptiles etc. through a religious medium. In the 'Tharavad' the Individual get the preliminary lessons about the entire ecosystem elements through a rigid religious and cultural framework. The institution of 'Sarpakav' is a sparkling example.

'Sarpakav'*

'Sarpakav' is an integral part of every 'Tharavad' system and is a strictly undisturbed ecosystem consisting of different types of trees (particularly, Kanjiram, Pala, Arayal etc.) creepers, medicinal and other plants, butterflies and other invertebrates, serpants and other reptiles. 'Sarpakav' being a religious institution, a belief is created among the members of the Tharavad from the very childhood that disturbing any element (say, plants, trees, butterflies, reptiles etc.) of the Sarpakav will effect them very severely through the wrath and displeasure of the God. At the same time, it is taught to them that watering the trees and plants in the 'Sarpakav' and giving milk to the serpents will give them life long prosperity and happiness (particularly by the women folks for getting off-spring). This religious adherence helped to maintain ecosystem to a very greater extent even today when there exists an extreme scarcity of land in Kerala.

'Temple'

Another ecologically oriented religious and cultural institution is the temples. Perhaps it is only in Kerala with each family (Tharavad) a temple is maintained. Temple, another integral part of 'Tharavad' system consists of trees,

*(Sarpa+Kav=Sarpakav, Sarpa=Serpant, Kav=An integrated Ecosystem consisting of all type of flora and fauna reserved for reptilse like a small National Park. May be the modern concept of National Park is attributed to the thousands of years old institutions of 'Sarpakav in Kerala).

plants, pond, animals (mainly cow and elephant) birds and a built structure and the deity. Here, again the individual is being taught from the very childhood that no ecosystem element like plants, trees, birds, and fishes in the pond of the temple system should be disturbed and fostering of these elements will bring the blessings of the presiding Deity/God. It is also made imprint in their minds that preserving and promoting the ecosystem elements of the temple will bring them prosperity through agriculture and feeding temple birds and animals will give brilliant offsrings.

The Institution of fostering of Rat-Snake

Yet another interesting religious belief imprinted in the individual is, fostering of 'Manja-chera' (Rat-snake) will bring fortunes. (It is logical here to drive that the maintenance of the Manja-chera (Rat snake) naturally brings down the population of rodents and thus avoid financial loss through crop damage caused by rodents and also brings down the incidence of plague.

The concept of planting and fostering of Mango trees and 'Tulasi' Plants

The importance of mango trees is so great that in each Tharavad mango trees are planted on all the corners of the four blocks of the "Nalukettu"[3] particularly on the immediate courtyard of 'Kizhakkini'—eastern block-under which a 'Tulasi' Thara (the platform where the 'Tulasi'[4] plants are planted) is erected, and an oil lamp is lighted every evening and prayer is offered to God for material prosperity, peace happiness and betterment of offsprings. Again it is here, i.e., under the mango tree, facing towards east the daily meditation is performed. The fruits of this particular mango tree is considered very sacred and often gives to patients with a belief that it has got a divine effect to cure ailments. The child in a 'Tharavad' is made to plant three mango trees with a view to attain 'moksha' and also material fulfilment during his life time as one mango tree to give him fruits, one for giving fruits to his offsprings, and another one for his funeral fuel/pyre. These all show not only the religious reverence of the people towards nature, but also the inventerate interelationship between religion, culture and ecosystem.

Ecological training through religious/cultural medium

The ecology oriented religious and cultural training the individual gets from the 'Tharavad' are many. For examples, the individuals are made to believe that unless they plant at least one account tree after the 'Vav Bali' (a religious rite done every year around the last week of July for bringing peace to the departed souls of the ancestors) the souls of their departed ancestors will not be at full peace. Again it is interesting that when an individual dies, to get 'Moksha' (ultimate peace) for his soul and also for his memory a coconut tree ('Chudala Thengu'[5]) is planted on the cremation bed. (No other place in India or in the world, such a great ecologically oriented religious practice is observed. When concrete structures are erected which are against ecosystem, a coconut tree is planted keeping in perfect harmony with the nutrient cycles of the ecosystem).

On the 'Pathamudayam' day (the 10th day of the Malayalam month Medam i.e., 23rd April, all the members of the Tharavad plant trees particularly, 'Phalavriksham' (fruit trees) believing that such a planting on that particular day will give them long life and prosperity.

On 'Onam' day (the annual religious Celebration day) the members of the Tharavad are strictly instructed not to distrub any plant or tree, birds, animals or other invertebrates (rather all living organisms).

The worship of Pipel tree

The worship of pipel trees, the unique cultural and religious manifestation of Dravidians, is strictly observed in Kerala. To get offspring, prayer and poojas are offered to pipel tree during 'Pournamies' (full moon day) and particularly on 'Buddha pournamy' day (the full moon day on which Lord Buddha was born). It is believed that making seven rounds around pipel tree and watering it on every day morning by the wife will bring immense prosperity and long life to the husband and children. Lighting the oil lamp on the bottom of the pipel tree, Kanjiram, Arayal, Pala (all tropical trees) on every evening is a features in Kerala and it is considered that such a deed will satisfy the Godess of Trees which in turn blesses with material prosperity.

It is strictly advised by the elders that no tree should be cut down but whenever a necessity arises the branches of the trees are only allowed to cut and not the very tree itself. However, when a tree is compelled to be cut down a prayer is offered to the 'Vriksha Devatha' (Godess of Trees) that the man responsible for cutting the tree may be forgiven and a new tree is planted by him to replace the one that was cut down.

All these religious and cultural practices shows that it is the religious and cultural factors which are the basic determining elements that help maintain the ecological balance in an enduring fashion and for such an ecology oriented religious and cultural attitude developed and inspired in the individuals, the family environment or the 'Tharavad' is the main responsible factor.

Conclusion

Thus, the Socio-cultural and religious practices of the Kerala Society is very unique and perhaps the most ecologically compatible way of life in the world. A perfect synthesis of Dravidian and Aryan culture, the Kerala Society manifest itself that an enduring ecological balance is more depending upon man's cultural and religious attitude and training than his economic exploitative attitude towards nature, as the former leads towards the latter.

This area of investigation in human ecology, religion and philosophy we have initiated through this attempt will have to be extended to other societies living in different parts of the world and we believe that such an attempt will go a long way in attaining complete harmony between man and nature. We also believe that in developing the 'preventive' technique for maintaining ecological/environmental balance these types of research attempts by social and behavioural scientists projecting the versatality and resiltience of social sciences in finding a permanent solution for environmental/ecological imbalance (rather than searching for solutions only within the premises of Science and Technology) is the need of the time.

Let What I dig from thee,
Oh, Earth, rapidly spring and grow again,
Oh, Purifier, Let me not pierce
through thy vitals or thy heart.

(Atharvaveda 12-1-1-35)

Acknowledgement

The author is indebted to Prof. P.R. Pisharoty for his valuable suggestions, help and encouragement. The author is thankful to him.

References

1. B. C. Sinha, Serpant Worship in India (1979), p. 15
2. B. C. Sinha, Tree worship in India (1978), p. 70
3. 'Nalukettu'=Nalu (Four)+Kettu (a built structure, a house) =a unique Pagoda type architecture system of Kerala where four houses are built together each facing East, North, West, and South and making a spacious courtyard) in the middle of the total structure of the four houses.
4. 'Tulasi'=A plant having medicinal and religious importance.
5. 'Chudala' (the Altar where the dead body is cremated)+Tengu =(coconut tree)=Chudala Thengo=A coconut tree which grows in the ashes of the Altar.

Acknowledgement

References

7

Ecological Consequences of Tehri Dam

*T. Shivaji Rao**

In their rat race to make money at any cost, vested interests behind development projects at the international and national levels are trying to force Governments in most of the developing countries to take up major industrial and irrigation projects that can be counter-productive in the long run. Politicians and people are made to believe that larger the project, greater will be the cost efficiency and social benefits. In countries like India as some politicians depend upon the financial support from the contractors to win elections and occupy seats of power, they via with each other to get as many projects as possible approved in their own states by the planning commission which sanctions projects that are socially justifiable and economically feasible. In the case of irrigation and power projects the increase in the number of major dams led to the dilution in technical vigour and impartial environmental impact assessment based on ecologically sound principles. The initial

***Professor, Centre for Environmental Planning, Andhra University, Waltaire.**

dedication to national cause among the politicians, bureaurats, engineers and contractors has given way to impatience and profit making at any cost. Although, the traditional financial productivity criterion for selection of projects was replaced by the "benefit-cost ratio", most of the time the exercise became redundant because the project authorities have recently made it a practice to over-estimate "benefits" and under-estimate "costs" to meet the statutory provision of establishing a benefit to cost ratio of $1\frac{1}{2}:1$. Because of such manipulations the cost over runs in the case of many dams constructed after independence were in the range of 100% to 200% and the underestimation of siltation rates in case of many dams led to a substantial reduction in their useful life ! As similar manipulations now being made in the case of Tehri Dam by the engineers, officials and Ministers are bound to cause serious disastrous consequences to the life and culture of U.P. in particular and the country in general, an attempt is made here to start a public debate on the prons and cons of the project.

Audit Objections

Although the local people were agitating against the Tehri Dam since 1969, the first national authority who fired the first bomb-shell against the economic and safety aspects of the project was Mr. T.N. Chaturvedi, the controller and Auditor General of India (CAG). In this audit report dated 22.11.1988, he clearly stated that experts gave different figures of seismicity for the design of the dam, while the Director of the National Geophysical Research Institute recommended in 1984 that the dam be designed for a horizontal acceleration of 0.56 g (g=acceleration due to gravity 980 Cm/Sec^2) due to an earth-quake of over 8 in magnitude on the Richter scale, the Roorkee University experts recommended in 1985 a safety factor of 0 25 g due to an earth-quake of magnitude 7.0. As the design of the project in 1986 was based on a safety factor of 0.15 g, the proponents stated in February, 1988 that the design will be worked out for seismicity of a higher order. The CAG warned that any subsequent change in the design of the dam based on a higher safety factor would considerably escalate the cost of the dam and adversely affect its financial viability. He also stated that the siltation rate adopted for the project seems to be on the lower side, affecting its life and financial viability adversely. The project was cleared by the Planning Commission in 1972 for an estimated cost of Rs. 198 crores to be completed in ten years to yield 325 MW of firm power and irrigate 6.67 lakh hectares at a cost benefit ratio of 1 : 12. The revised project estimated at Rs. 3000 crores provides irrigation facilities to 2.7 lakh hectares and generates power of 2000 MW at Tehri and 400 MW at Koteshwar dams at a cost benefit ratio of 1 : 3½ as against the stipulation of 1 : 1½ made by the Planning Commission. The project works which hitherto had been under the Control of Government of U.P. were transferred on 1.6.89 to the Tehri Hydro Development Corpn. Ltd., a Company incorporated on 12.7.88 as a joint venture of the Government of India and the Government of Uttar Pradesh. The project is now being executed with the technical and financial assistance upto Rs. 1800 crores from Russia. Due to persistent opposition by the local people to the Tehri project Mrs. Indira Gandhi discussed the damaging environmental impact of the dam with the proponents and opponents and realised that there is great local distress and a feeling that contractors and other such groups will be the main gainers and hence ordered in March, 1980 for a review of the projects such as silent valley in Kerala, Tehri in U.P. and Lalpur in Gujarat from the environmental angle. The working group constituted for environmental impact of Tehri was headed by Sri S.K Roy who expressed his strong opposition to the dam in his report submitted to the Ministry of Environment on 28.10.1986. Although the Vice-chancellor of Roorkee University, the Chief Engineer, Tehri Dam Project and the Chief Engineer of Irrigation Department U.P. wrote letters to the Ministry of Environment to ignore the final report submitted by Sri Roy, the Union Ministry recommended for abandoning the Tehri Project. On 16.10.1986 the Central Water Power Commission discussed the feasibility of construction of Tehri dam from seismic considerations and decided that as recommended by Roorkee experts the effective peak ground acceleration at Tehri dam site will not exceed those corresponding to an event of magnitude 7.25 due to major faults and the effective peak ground acceleration of 0.25g for the maximum credible earth-quake was considered adequate to ensure safe design of the dam. As a consequence, the Government entered into an agreement with the Soviet Union in November, 1986 for technical and financial assistance. The rated accelerogram suggested by the Indian experts with the maximum acceleration of 0.45 g was accepted by the Soviet experts and the average slopes of 1 : 2 for down-stream and 1 : 2½ for up-stream sides were fixed for the rock-fill dam with side shells of sandy gravels. During January 1987, the Ministry of Environment was forced by the vested interests to give a press release, stating that "the Government have cleared the project after a through assessment of the impact of the project on environment and after satisfying themselves that the adverse impacts on environment can be remedied". Meanwhile, the Tehri Virodhi Sangarsh Samithi filed a petition in the Supreme Court in 1985 pleading that the Tehri Project should be abandoned because of its economic non-viability and inherent danger to the life and property of lakhs of people.

Warnings by experts

When the project authorities asked for funds for the work, the Union Ministry of Finance stated in July, 1989 that further release of funds will be made after securing the usual environmental clearance. Accordingly the authorities submitted the environmental action plans on 29.11.89 and the same were first discussed by the standing committee on environmental appraisal of River Valley and Hydro-Electric Projects of the Government. After a field visit to the project site and prolonged discussions with the project authorities, experts and others, the standing committee of environmental experts headed by Dr. Bhumbla recommended unanimously on 15.2.90 for abandoning the project on the grounds of seismic potential, risks due to failure of dam, disaster management, siltation, land slides, and other environmental factors. The committee warned that if the dam collapsed or was overtopped due to collapse of the hill-sides due to an earth-quake of over 8 magnitude on the Richter scale, the resultant water wave would wipe out Rishikesh, Hardwar and their environs. Since the scientific assessment of Tehri project by Bhumbla Committee was unpalatable to the vested interests, the Prime Minister was unduly pressurised to appoint a 5-member non-environmental experts committee headed by Dhoundial to examine all safety and seismological aspects of the project. It was amply advertised that the Dhoundial Committee comprises foremost experts in the country on the subject of seismicity and design of dams in earth-quake prone areas and that the members also represent at the highest level the organisations dealing with these issues. The Dhoundial Committee examined the Bhumbla Committee report and accepted the recommendation that Tehri site earthquake potential has to be revised from 7.25 magnitude to 8.5 magnitude. (Appendix-I). The experts committee headed by Bhumbla made an indepth study on siltation, landslides, seismicity, dam-burst and risk analysis as also the feasibility for disaster management and ultimately concluded that clearance for the Tehri project as proposed should not be given. It was estimated by the Committee Members that the costs of damage to life culture and properties as consequence of the inevitable dam burst will be for higher than the economic benefits sought to be obtained from the project.

Based upon a highly unscientific assessment on the seismic potential and safety aspects of Tehri Dam that was objected by Dr. V.K. Gaur, one of the Members of the Committee, a conditional environmental clearance was granted under unusual circumstances by the Union Ministry of Environment for the Tehri Dam Project. The first condition stipulated by the Government States that the Tehri Development Corpn. Ltd. must get the safety aspects and design of the dam approved by the High Level Expert Committee constituted for the purpose and the second condition stipulates that Comprehensive Environmental Management Plans must be formulated to the satisfaction of and got approved from the Ministry of Environment in a time bound manner as outlined in their communication, No. 2-19/81-HCT/IA-(I) dt. 19-7-90.

Dam Safety Doubted

By imposing the first condition, the Government has indirectly accepted the recommendation of the 13-member Environmental Appraisal Committee of Ministry of Environment headed by BHUMBLA that the safety of the dam is very doubtful. The Government could not perhaps accept fully the recommendation of the 5-member High Level non-environmental Expert Committee headed by DHOUNDIAL, an expert in Geology because one of its members Dr. V.K Gaur is reported to have raised an objection about the safety of the dam on the basis of new information on an important design aspect obtained from an international expert Prof. JAMES BRUNE of USA that the maximum peak ground acceleration (PGA) at Tehri site (a crucial design parameter) will be about 1.0 g ($g=980$ cm/sec^2 the acceleration due to gravity) and not 0.44 g as recommended by the local civil engineers of the expert committee. It is true that the consulting civil engineers who insisted that Tehri site will not experience an earth quake of more than 7 magnitude on the Richter Scale for a number of years upto Feb '90 ultimately recognised the true scientific aspects of seismicity changed their stand and confirmed in April '90 that an earth quake of 8.5 magnitude will occur at Tehri site during the life-time of the dam as recommended by the Bhumbla Committee. The Dhoundial committee confirmed the focal depth of the earthquake as 15 KM hence the hydro-central distance will be 18 to 23 KM from the project site.

Seismicity Revised

According to Soviet expert SHEBALIN'S relation between intensity, I, magnitude, M & hypo-central distance Hd, $I=(3+1.5\ M-3.5 \log Hd)$ it is found that an earth quake of magnitude over 8+on Richter Scale exactly corresponds to intensity of about 12 on the Modified Mercalli Scale.

According to JAMES L. SHERARD, an internation expert on Earth and Rock-fill dams, it must be assumed that all dams located in any large general areas of seismic activity such as the state of California may be equally severely shaken regardless of their location within the area. Consequently Tehri site which is in proximity to Kangra that experienced an earth quake of magnitude 8.6 on 4-4-1905 must be assumed to experience an earth quake of the same magnitude of 8.5 within the life-time of the project and since the focal depth is taken as 15 KM by the experts, the intensity is bound to be 12 on Modified Mercalli scale as in the case of the Meghalaya earth quake of 12-1-1897 and Eastern Arunachal earth quake of 15-8-1950.

An earth-quake of magnitude seven (intensity 9) is equivalent to an explosion of about 90,000 tonnes of TNT while that of 8.5 magnitude (intensity of 12 points on the Modified Mercalli Scale) is equal in explosive power of 28.6 million tonnes of TNT. A dam designed to withstand a given level of energy of explosion cannot be expected to remain intact when exposed to 300 tonnes higher levels of explosion and that too in the tectonic boundary of the most destructive subduction zone of the Alphide-Himalayan Seismic belt with a foxal depth of 15 KM. Hence the contention of the Dhoundial Committee that since effective peak ground acceleration adopted in the design was 0.25 g the present design of the dam would be capable of withstanding even an earth-quake of magnitude 8.5 becomes invalid.

Manipulations

The effective peak ground acceleration (PGA) of 0.25 g is based on the maximum acceleration of 0.45 g. deduced from the MC GUIRE relation between acceleration, distance and magnitude of Seven on the Richter Scale (equivalent to intensity 9). By revising the magnitude of the earth-quake from 7 to 8.5 magnitude on the Richter Scale, the Dhoundial Committee confirmed that the intensity has been correspondingly increased from 9 to 12 points on the Modified Mercalli or the Russian Scale. But the experts instead of using 8.5 in place of 7 in MC. Guire equation conveniently adopted a less reliable method.

For the earth-quake magnitude of about 8.5 the PGA according to MC. GUIRE equation works out to many times the acceleration due to gravity and disqualifies the site for the dam. Hence the experts suddenly abandoned this method and opted for a method on accelerations in the rocks due to earth-quakes. On the basis of strength of most rocks, when HOUSNER a U.S. expert contended that according to the concept of an upper bound to maximum acceleration, the maximum acceleration cannot exceed 0.5 g due to an earth quake from a fault stippage, other experts like NEWMARK and ROSENBLUETH instead on raising the acceleration to at least 1.0 g and perhaps even 1.5 g in the light of reports on high accelerations experienced during the Assam Earth quake of 1897. A international expert D.J. DOWRICK in his manual on earth-quake resistant design presented another attenuation relations deducted by another internation expert ESTEVA on the basis of Californian data on accelerations in rocks and this equation gives accelerations of 450, 670, 1000 and 1500 cm/sec^2 for magnitudes 7, 7, 5, 8 and 8.5 respectively and they correspond to about 9, 10, 11 and 12 intensities for Tehri. These accelerations are also in agreement with accelerations presented graphically by Dr. Jai Krishna, an expert of Roorkee, in a paper submitted to a seminar on Geophyscial investigations in the Peninsular shield. Since the application of these established scientific, technical and environmental principals have been given up by the Dhoundial Committee and the Secretaries to Government manipulations by experts became inevitable to justify a predetermined conclusion that Tehri Dam will be absolutely safe.

According to Prof. N.N. MASLOV a reputed soviet expert, when increasing earth-quake intensities by 1 force number, it is necessary to increase two-fold the design values of seismic acceleration and consequently the seismic constant (coefficient). The relationships between earthquake intensities and peak ground accelerations (PGA) deduced by international experts like RICHTER, NEUMANN, MEDVEDEV, TRIFUNAIC and MURPHY are in conformity with this principle. According to the MEXICAN expert Prof. LEDNARDO

ZEEVAERT the accelerations of ground surface deduced by RICHTER after a careful study of californian data for intensities of 8, 9, 10, 11 and 12 are 150, 320, 675, 1500 and 3200 cm/sec^2 respectively.

But the Dhoundial Committee virtually ignored this fundamental scientific principle. According to Prof. MASLOV for the earth-quake intensities of 8, 9, and 10 the design seismic accelerations are 100 to 200 cm/sec^2, 200 to 400 cms/sec^2 and 400 to 800 cm/sec^2 respectively. According to the American experts MURPHY, TRIFUNAC and NEUMANN who independently worked out the calculations, peak ground accelerations for intensities of 8.9 and 10 are in the ranges of 230 to 265 cm/sec^2, 517 to 538 cm/sec^2 and 1033 to 1160 cm/sec^2 respectively.

It is well-known that when a major earth quake occurs in the seismic gap of the Himalayas, the rupture along the plane of under-thrusting of the Indian plate, whose shortest distance from the damsite is 15 km, would be the main source of high peak ground acceleration (PGA) at the Tehri dam site. The attenuation factor of 0.223 over a distance of 15 km, thick rock material is estimated by Dhoundial committee to produce a peak acceleration of 0.446 g by applying the formulation presented by Prof. James. N. Brune of U.S.A. in his publication on "The physics of Earth Quake strong motions" in 1976. When requested to clarify whether his formulation has been properly used by the Dhoundial committee in arriving at the relevant peak ground acceleration expected at Tehri due to a major earth quake of over 8+magnitude anticipated in the project area, Prof. Brune replied that Dhoundial has misused his formulations. He emphasised that since Tehri dam is located in the world's greatest zone of thrust-faulting and that too in a seismic gap as well as a geological structure strongly suspected of producing accelerations over 1 g as reflected by the Assam earth quake of 1897, the peak ground acceleration at Tehri will be more than 1 g. In fact, stress drops in the thrust zones being higher than the average, high accelerations of over 1 g have been produced by at least 3

Table 1

Numerical Values for Some Suggested Relationships Between Peak Ground Acceleration and Modified Mercalli Intensity, or Equivalent*

Mod. Mercalli Int. or equivalent	*Hershberger (1956) Ave. Accel.*	*Richter (1958) Ave. Accel.*	*Neumann (1954)*	*Medvedev and Sponheuer (1969)*	*Japan Meterological Agency (Okamoto, 1973)*	*Trifunac Hariz. Ave. Accel.*	*Trifunac Vert. Ave. Accel.*
VII	128.8	64.	130.	50-100	44-94	126.	83.
VIII	346.7	138.	265.	100-200	94-202	251.	166.
IX	933.3	295.	538.	200-400	202-432	501.	331.
X	2512.	631.	1094.	400-800		1000.	676.
XI							
XII							

*Acceleration is measured in centimeters per second per second.

thrust-fault earth-quakes (San Fernando, 1971; Gazli, U.S.S.R., 1976 and Nahinni, Canada, 1985). Even if the effect of local amplication of acceleration caused by topographical irregularities is neglected for Tehri Project, the free surface amplification by a factor of two and the surface effect of low velocities may amplify ground motion, resulting in a peak ground acceleration in excess of 1 g. Brune emphasised that a dam in similar circumstances as Tehri Dam could not be licenced for 8.22 g in U.S.A. He found that experts consider 0.22 g as too low a design acceleration for Tehri dam.

According to the relation between magnitude, distance and acceleration graphically presented at the second symposium on Earth Quake Engineering in 1962 by Dr. Jai Krishna of Roorkee University an acceleration over 1 g occurs at Tehri unfortunately, he subsequently slipped into a wrong track in March, 1969 on the basis of the unreliable presumptions of Housner of U.S.A., in fixing a ceiling limit of 0.65 g for the peak ground accelerations while actual field observations in different countries during major earth quakes proved that such arbitrary limits on accelerations are not only unscientific but also hazardous and devastating in their damaging impact on the environment. Dr. Bharat Singh of Roorkee University ultimately accepted the graphical presentation made by Dr. Jai Krishna in 1962 and recommended its use in his classical text book on "Earth and Rock-fill dams in 1976. The Dhoundial committee stated that under the inter-governmental agreement between the Government of India and the Government of U.S.S.R. for cooperation on Tehri Project, the section of the dam as designed by the Indian Engineers stands validated by an independent review as per U.S.S.R. practice. It is a pity that the Indian experts designed the dam for an intensity of 8 and the Soviets revised the seismic potential to intensity of 9. But the Indian experts who designed the dam for magnitute 7.2 on Richter scale in 1989 have not perhaps informed the Soviet experts so far that the Seismic potential has been revised by them to 8.5 on Richter Scale or 12 on the Modified Mercalli scale. According to Russian Norms, Seismic potential is indicated in terms of intensity and peak ground accelerations are based on intensity scale. Since the Soviets reject sites with seismic potential above nine intensity Tehri Dam site will be rejected by them (see APPENDIX-II).

Tehri Dam Unsafe

In order to find out whether the Tehri dam as presently designed is safe or not, one can use the basic principles of engineering machanics and the seismic coefficients as per Russian norms cited. In fact the U.S. Army Corps. of Engineers suggested in their design manual, the use of an effective slope of equivalent slope for seismic forces. Consider a slope of inclination 'i', and a plane at a vertical depth, H, below the surface of the dam and take an element of unit width as shown in the figure.

Weight of the element = W, its normal component = W cos i, and its tangential component = W sin i, As the area of the inclined base of element $= \frac{1}{\text{Cos i}}$ the intensity of normal Stress $= (\text{W Cos i}) \div \left(\frac{1}{\text{Cos i}}\right) = \text{W Cos}^2 \text{ i}$ and the intensity of tangential Stress = W sin i Cos i.

Now, let the elen ent be subjected to a horizontal inertial force due to an earth quake of "α W" (α. g. = horizontal acceleration) This causes a down slope tangential stress "α W sin i : cos i".

For equilibrium, the tangential stress should not exceed "Tan ϕ" times the normal stress, where "ϕ" is the angle of internal friction for sandy gravel used for the construction of the dam and "i" is the slope angle; therefore, (W Sin i. cos i + α. W. cos^2 i) = tan ϕ (W cos^2 i = α. W sin i cos i)

Divide both sides by "W sin i, cos i", then 1 + cot i = (cot i $-$ α)/cot ϕ.

Let "Cot ϕ" which is the stable slope without the earth-quake be represented by "M′" and "Cot i" is the just stable slope with the earth-quake be represented "M".

Now, $1+\alpha.\, M=\frac{M-\alpha}{M'}$

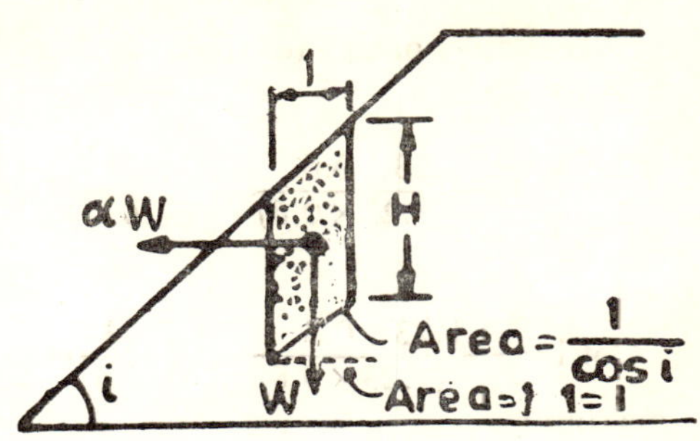

$M'+\alpha.\quad MM'=M-\alpha$

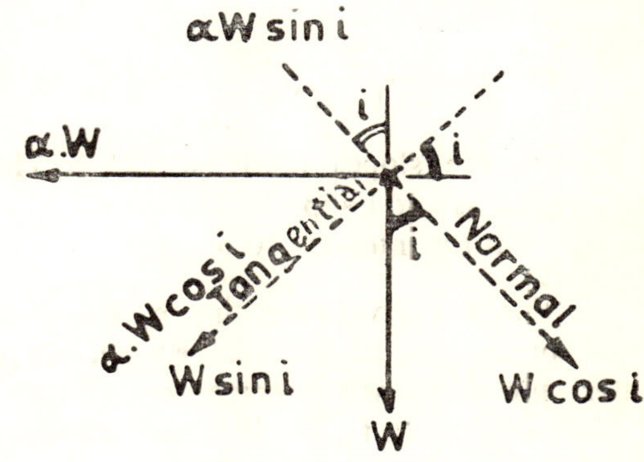

$M'+\phi=M-\alpha\ MM'=M\ (1-\alpha\ M')$

$$M=\left(\frac{M'+\alpha}{1-\alpha\ M'}\right)$$

For sandy gravel used, $\phi=39°$; Cot $\phi=1.23$; Seismic coefficient=0.25 as recommended by the experts of ROORKEE

$$\therefore \qquad \text{Cot } i=\frac{1.23+0.25}{1-(1.23\times 0.25)}=2.2$$

Accordingly the slopes of the dam are fixed as $2\frac{1}{2}$: 1 & 2 : 1, However, if the average of the design acceleration for the range 400 to 800 cm/sec^2 for the intensity 10 on Russian Scale is used as per Russian norms, ($\alpha=0.6$) as also the max. of design acceleration $\alpha=0.8$.

$$\text{Cot } i=\frac{1.23+0.6}{1-(1.23\times 0.6)}=\frac{1.83}{0.26}=7.4;$$
(average)

$$\text{Cot } i=\frac{1.23+0.8}{1-(1.23\times 0.8)}=\frac{2.03}{0.016}=127;$$
(maximum)

Accordingly the slopes of the dam must be 8 : 1; for $\alpha=0.6$; and 127 : 1; for $\alpha=0.8$; By using RICHTER'S and ESTEVA'S accelerations for intensity ten as cited by ZEEVAERT and DOWRIC

$$\text{Cot } i=\frac{1.23+0.67}{1-(1.23\times 0.67)}=\frac{1.90}{0.17}=11.2$$

Accordingly the slopes must be 12 : 1; The mountain slopes will collapse as had happpened at Vajont Dam in Italy in October, 1963 at the intensity 10 or more.

As per this conservative estimate, the Tehri dam as designed is bound to collapse. Otherwise, if the slopes are made flatter, the dam will become too costly. Hence the Russians reject sites prone to earth quake of intensity more than 9 for building major dams in their own country.

DAVIDOV the Russian expert has already stated in February, '90 that the Tehri dam can withstand an earth quake of intensity 9 only and than ten points intensity earth quake is improbable at Tehri project site. Since the Dhoundial committee has confirmed that Tehri site will experience an earth quake of 8.5 magnitude (equivalent to 12 points intensity on Russian scale), the question of providing the safety of the Tehri dam as designed in its present location as per the rigorous scientific and technical norms becomes a day-dream and hence the conditional environmental clearance accorded to the Tehri dam tantamounts to virtual rejection of the Tehri dam. Although the Ministry of Environment gave a similar clearance to the Tehri dam by its press notification in January '87 the Union Ministry of Finance insisted in July '89 that further release of funds for the dam will be considered if only a clear approval is given to the project from the environmental angle. Since the environmental clearance is not a clean chit but sustains public objections against dam safety and environmental risk and is subjected to so many critical conditions that cannot be fulfilled, the Union Ministry of Finance cannot release any further funds to the project particularly when the environmental clearance to the Tehri dam becomes a virtual negation of the Project itself.

Safety Norms Violated

When the central Water Power Commission suggested a seismic coefficient of 0.15, and the Rookee experts 0.25, the USSR experts recognised the defective design and recommended for a dam design that withstands an earth quake of a 9 intensity. Dr. JAMES L SHERARD warns that the practice of providing a safety factor of about 1.1 under a static horizontal force 10% to 15% of gravity gives no assurance that the dam will be safe under strong earth-quake shaking. Since the ground shaking motion in a strong earth quake has acceleration pulses of 50% of gravity, the analytical procedures cannot give any confidence that large shearing movements could not occur. Based upon his thirty years of professional practice in the most seismic area of Mexico city Prof. ZEEVAERT the consultating civil engineer of international repute concluded that seismic design for 9 intensity is the maximum possible for engineering purposes since intensity ten is considered catastrophic as the ground would not support construction in a satisfactory way, even if the construction could be designed for an earth quake of such magnitude. According to the world famous Soviet expert on dams Dr. M.M. GRISHIN, Rock-fill dams are constructed in areas of high magnitude earth-quakes upto nine intensity only.

According to S.V. POLYAKOV another Soviet expert the usual anti-seismic design measures prove inefficient in the cases of intensity ten and more violent earth-quakes and that is why the new Soviet intensity scale contains no relevant data on ground accelerations, PGA. According to the Soviet norms, Tehri Dam project site has to be abandoned as it will experience earth quake of intensity of 12 and the dam will collapse wiping out many pilgrim towns like Rishikesh, Hardwar and Deoprayag and villages of U.P. resulting in heavy loss of life and properties worth several thousand crores !

It must be remembered that the Soviet expert Mr. FINK in his rejoinder to a news item in the Times of India (dt. 15-11-88) entitled "USSR. Not keen on Tehri project" confirmed that they had kept in view the seismicity of Tehri project region as nine points on the Modified Mercalli Scale. As the Russians were mislead into believing that the Seismic potential of Tehri site is exactly similar to the site selected for Nurek dam (300 meters high) in USSR the soviet Govt. agreed to offer both technical assistance and financial assistance of Rs. 1,800 crores as also to participate in the construction of the project works. Now that the Dhoundial committee has belatedly realised the scientific truth that Tehri site will be exposed to an earth quake of intensity 12 on the Modified Mercalli scale, the Soviet experts in Moscow must be requested to reassess the technical feasibility of the project and the safety of the dams as per revised seismic potential. If the Soviet experts are to adopt the highest safety norms the lives of Indian people and their culture for the design of the Tehri dam, they will automatically advise both the Russian and Indian Governments to abandon the project from the angles of dam safety and environmental risks. Thus Tehri site is not acceptable for a major dam construction either as per Russian or American safety standards. So, it has to be rejected in its present form. However, run-of-the river schemes as recomemded by the Bhumbla committee may be executed for minimising the economic loss to the nation and the same economic of aims sought by the Tehri dam project can be secured by several other alternate methods.

Dam Disasters

Surprisingly Dhondial who expressed serious concern about the environmental hazards due to gigantic land slides at Nainital (September, 1880), the two rock falls that plugged the river course, forming about 275 meter high dam around Gohna lake (September, 1893) and the Belakuchi disaster (July, 1970) at a seminar in New Delhi on 14-11-1977 miserably failed to warn the Government that the Tehri dam is bound to fail one day or the other either due to an earth-quake of magnitude over 8 or due to massive land slides from the steep hill slopes around the reservoir. In fact the Kunnaudiaged land slide (1978) caused havoc in the Gangetic valley. The Vajont dam in Italy failed (October, 1963) not because of any structural defect but due to the collapse of the mountain slopes around the reservoir. The Machchu dam in Gujarat with 50 feet water depth in the reservoir burst on 11-8-1979 when the flash floods reached Morvi town 5 km. downstream in 9 minutes killed thousands of people and destroyed the industries and buildings by a 30 feet high battering ram of water. The panshet and Khadakwasla dams burst in 1962 in Maharashtra causing wide spread damage and loss of life. Due to incessant rains in U.P. state Nanak Sagar and Sarada Sagar were about to collapse on 6-8-1980 when human and animal populations in about thousand villages were threatened to be washed away. Despite the assurances given to the environmentalists by the engineers and the Government about safety, the Vajont dam in Italy failed in October, 1963 as anticipated by the ecologists and the local people. In a technical paper on the flood surges due to the failure of Tehri dam before the Institution of Engineers, Visakhapatnam on 5-6-90, Dr. Narasimham estimated the flood waters reach Rishikesh in an hour after the dam burst and Hardwar in another 20 minutes and wipeout Deoprayag, Rishikesh, Haradwar and their environs by more than 200 feet high battering ram of water. The results of such failure result in deaths of millions of jat farmers poor muslims and destruction of Roads, Railways and irrigntion systems in U.P. worth about Rs. 20,000 cros.re

When dam-break analysis reports and inundation maps for evacuation and disaster management are produced by British Columbia Hydro in Canada, why should the Government fail to prepare such reports for Tehri before clearing the project ?

Accordlng to the modern theory of plate tectonics, the Indian plate is penetrating under the Asian plate in the Alphide—Himalayan seismic belt. Due to the specific local features of the upper crust and soils the seismic waves in this belt of subduction produce higher accelerations and intensities. For example the 6.8 magnitude earth-quake that rocked Kinnaur on 19-1-1975 with a focal depth of 29 km. produced an intensity of 9 on modified Mercalli scale while an earth-quake of magnitude 8.7 with a focal depth of 30 to 50 km. in Eastern Arunachal Pradesh produced an intensity of 12 on MM scale on 15-8-1950. Hence an earth-quake of over 8 magnitude 6.4 with a focal depth of 14 KM at Koyna produced an intensity of 8 with peak ground acceleration of 0.6 g on 10-12-1967. While an earth-quake of magnitude 7.5 killed one lakh people in Italy in 1908, a Quake of magnitude 8, killed 2,40,000 people in China in 1976. The latest earth-quake of Magnitude 7.3 killed about a lakh of people in Iran and destroyed Raksht dam and irrigation systems and properties in June, 1990. Since no dam can be cent percent fail-proof the basic inundation map should have been prepared in the first instance to find the feasibility for emergency evacuation and rehabilitation including payment of compensation for lives and properties likely to be affected due to failure of the dam.

Evacuate People Upto New Delhi

Since the people of U.P., Bihar, West Bengal and perhaps Bangladesh will be affected consequent to the failure of the dams due to violation of all the norms in siting, design and environmental protection, the people must know whether the U.P. state Government or the Central Government or the

Soviet Government involved in the construction of these hazardous dams will be prepared to pay the compensation as determined according to international norms. While the dam safety Act, 1972 of U.S.A. insists on the dam break analysis and disaster management plans for clearing the sites for major projects, the Federal Act on dams in Switzerland provides compensation to victims and loss of properties due to dam failures from natural causes or enemy action. By branding the Tehri dam as a "fail-safe dam", the secretaries to Government of India are blatantly misleading the people and the Government because when the dam fails it does not remain safe but causes devastating floods that kill lakhs of jat farmers, millions of poor muslims and lakhs of Hindu pilgrims that frequent the sacred places like Hardwar for Worship during Kumbh Melas etc. It looks as if the project is designed to fail and if it can not be stopped at least people from all the villages and towns in Western U.P. and other places downstream are evacuated from the flood-prone zones in the Gangetic belt even before rehabilitating the people from the Tehri area. Since members of Dhondial Committee have vested interests, their report is naturally biased and misleading. Under the circumstances the scientists journalists and the educated people from all parts of the country must discuss about the pros and cons of the project and advise the Indian and Soviet Governments to modify the project after holding public debates on the crucial problems of the project as high-lighted in the reports of different expert Committees.

APPENDIX I

Contradictory Views on Tehri Dam

(a) Scientific Views of Expert

According to Prof. James N. Brune, the Himalayan region in the vicinity of Tehri dam is one of the most interesting on earth, for a number of reasons :

1. It is the only place on earth that two continents are crushing together at a high rate (about 5 cm/yr),

and have been doing so for tens of millions of years.

2. The Himalayas are the highestmountain range in the world, a consequence of the intense effects of the collision, which has caused uplift of rocks at such a rate that erosion could not wear them down as it would in many other areas of the world. The Himalayas produce the greatest amount of erosional products of any region on earth. These are delivered into the oceans,—the Arabian Sea and the Bay of Bengal, by the Ganga and Brahmaputra rivers, and by the Indus River, and produce the largest and most rapidly accumulating deposits of sediment in the world.

3. To keep the Indian plate moving into Asia as it has for tens of millions of years requires a tremendous expenditure of the energy which drives tectonic plates, and a tremendous amount of energy is stored elastically in the crust at any given time. There is every reason to believe that the available rupture energies in the colliding continental structures of India and Asia are as high as anywhere on earth.

4. The Himalayan thrust belt has in historic times produced some of the largest and most energetic earthquakes in the world. These great earthquakes can be expected to continue to occur indefinitely.

5. A major seismic gap exists between the rupture zones of the 1905 and 1935 earthquakes (the Central Gap of Khattri, et.al.), and this gap can be expected to be filled in the future with an earthquake of comparable great size. This gap includes the site of the proposed Tehri Dam. A potential slip of considerably more than 10 meters may have been stored up.

6. There is extreme danger not only from the major thrust fault, but also from numerous imbricate

branching thrust faults which might rupture the crust much nearer the dam than the main fault itself, possibly producing peaks of ground acceleration even greater than produced by the rupture on the main thrust.

7. The Himalayan thrust belt produced the first clear evidence that the ground acceleration exceeded the acceleration of gravity, and objects were actually thrown into the air. This was documented by Oldham in his study of the great 1897 Assam earthquake. Although this evidence was considered with skepticism for many years, with the advent of modern strong motion instruments which can reliably measure such strong accelerations, there is no longer any basis for skepticism, since we have recorded accelerations near and over 1 g for several large earthquakes.

Table 2

Earthquakes with Ground Motion Exceeding 1.0 g

Instrumental		
San Fernando, Calif., USA	1971	1.25 g
Gazli, USSR	1976	1.35 g
Imperial Valley, Calif., USA	1979	1.75 g
Victoria, B.C., Mexico	1987	1.0 g
Nahanni, N.W. Terr., Canada	1985	2.0 g
Loma Prieta, Cal. USA	1989	1.0 g
Observation of the Effects		
Assam, India (Oldham)	1897	1.0 g

Considering the above points, we have to conclude that the proposed Tehri dam location is one of the most hazardous in the world from the point of earthquakes.

According to Prof. Brune, there are a number of circumstances which suggest that accelerations near 1 g are likely for a major earthquake in the neighborhood of Tehri Dam :

1. Such high accelerations have occurred in an earlier Himalayan Thrust earthquake,--the great 1987 Assam earthquake.

2. The high stresses thought to be required to keep the Indian subcontinent moving into Asia might indicate proportionality larger amounts of energy stored in the crust of the earth waiting to be released during an earthquake.

3. There is currently debated evidence that continental thrust faults of the type associated with the Himalayan thrust are associated with larger stress changes (stress drops), and thus may generate higher ground accelerations than other types of earthquakes (e.g., strike slip earthquakes common in California, and normal faulting earthquakes common in Nevada).

4. Our current understanding of the effects of seismic wave attenuation, near-surface resonance and amplification, and the effects of focusing of energy by rupture propagation, all support the possibility of ground accelerations greater than 1 g from a large earthquake near Tehri dam.

5. Recent statistical predictions based on extrapolations from the current data base of observations of ground accelerations, suggest that for Magnitude 8% earthquakes, as expected for the Tehri dam region, accelerations of over 1 g may be expected approximetely 16% of the time.

(b) Unscientific Views of Bureaucrats

Contrary to the above scientific findings from an international expert, the Government of India was pressunised by the vested interests to reject the scientific findings of the standing committee of experts of the Union Ministry of Environment, headed by Bhumbla. For this specific purpose, a committee of secretaries to the Government was constituted in March 1990 to get a tailor-made report to justify continuation of work on Tehri dam. According to Johnsons committee report on "Impact analysis of the water resources planning and management" to the American Society of Civil Engineers envisaged under the National Environmental Policy Act of 1969, a water resource project will normally have a significant effect on the environment if it will (a) disrupt or alter archaelogical or historical sites (b) cause substantial flooding, erosion or siltation (c) expose people or structures to major geologic hazards and (d) create a potential public health hazard. In tune with the civil engineers' responsibility in impact analysis with a fundamental concern for the health, comfort and welfare of man, the engineers of the Bhumbla committee made an indepth study of all the relevant crucial environmental problems of the Tehri dam. As the secretaries to the Government were not experts in the field of environment, they could not appreciate the high scientific and technical inputs injected into the environmental appraisal of Tehri dam and found it convenient to byepass the Bhumbla committee report under the mischievous plea that the Bhumbla committee ought to have concerned itself with the environmental parameters within which the opinion of the said committee was relevant. It is surprising that the secretaries to Government were pressurised to develop wrong notions that environmental safety of the Tehri dam should be decided not by the environmental experts and Engineers of the Bhumbla Committee but by the non-environmenal experts of the Dhoundial Committee who have vested interest in the project.

APPENDIX-II

Intensities of Earth-Quakes And Their Effects

(In Russian Scale or Modified Mercalli Scale)

1. No damage to houses or structures. No residual disturbances in soils or groundwater and surface water regime. Earthquake not felt (classified as *Unnoticeable*) and only instrumentally recorded.

2. No damage or disturbance inflicted to structures. Felt only by a few persons and classified as *Very Feeble*.

3. No damage or disturbance of structures. Suspended objects, such as lamps, curtains, open doors slightly swing. Standing cars rocked inappreciably. Felt as vibrations of rapidly moving vehicles. Classified as *Weak*.

4. No damage in buildings or structures. Suspended objects swing, fluids in vessels oscillate, dishes disturbed. Felt by nearly every one. Classified as *Moderate*.

5. Damage slight in structures. Open doors and windows swing, window panes oscillate. Dishes and objects on shelves overturned. Felt by almost everybody; domestic animals agitated. Classified as *Fairly Strong*.

6. Minor damage in many buildings. Light furniture disturbed. Many run outdoors. People move unsteadily. Animals run out of shelters. Classified as *Strong*.

7. Damage slight to considerable in many structures. Masonry fences damaged. Roads cracked; landslides, occasionally rock slides in mountainous areas. Previously existing springs disappear new springs

emerge. Some persons jump out of windows. Motion without support made difficult. Classified as *Very Strong*.

8. Damage considerable in structures, some buildings collapse. Largescale talus slides and rock falls. Persons can hardly stand. Classified as *Destructive*.

9. Structures badly damaged, buildings fall. Fall of most chimneys, stacks, towers. Rails bent. Rock avalanches. Classified as *Destructive*.

10. Structures often collapse. Exbankments and dams badly damaged. Ground badly cracked by fissures 1 m. wide. Fall of Cliffs. Battering of waves of reservoirs. New lakes originate. Animals rush about. Old trees broken. The earthquake is classified as *annihilation*.

11. Few (if any) structures remain standing. Long sections of embankments collapse. All rails warped and buckled. Numerous fissures on ground, vertical displacement of soil strata. Classified as *Catastrophic*.

12. *Major Catastrophy*. Damage total. Large-scale changes of ground : overthrusts, ruptures, landslides etc. tremendous.

(From : "Basic Engineering Geology and soil mechanics" (1987) by N. N. Maslov, Page 505, Mir Publishers, Moscow)

References

1. Bharat Singh et. al.; "Earth and Rock-fill Dams", Sarita Prakashan, Meerut (1982).
2. Bhumbla Committee report on Environmental Appraisal of Tehri Dam Government of India, New Delhi (Feb, 1990).
3. Brune N., James, "The seismic Hazard at Tehri Dam", Paper from INTACH, New Delhi (Dec. 1990).
4. Chaturvedi, T.N., CAG Report on Tehri Dam, Government of U.P., Lucknow (1988).

5. Dhondial Committee report on Tehri Dam, Government of India, New Delhi (April 1990).
6. Dowrick, D.J., "Earth Quake Resistant Design", John Wiley, New York (1977).
7. Davidov, "The Hindu" Dated 25-4-1990 and, "Times of India" Dated 10-2-1990.
8. Fink "Times of India" (Nov. 1988).
9. Grishin, M.M., "Hydraulic Structures", Mir Publishers, Moscow (1987).
10. Jai Krishan., "Earth-Quake Resistant Design of Earth Dams", Second symposium on Earth Quake Engineering, Roorkee (1962).
11. Jai Krishna, "Ground accelerations during Earth-Quakes", Bull., Ind., Soc., Earth Quake Tech,, Roorkee (March 1969).
12. James., L., Sherard, "Earth-Quake Considerations in Earth Dam Design", Soil Mech., Division, Am., Soc., Civil Engineers, New York (July 1967).
13. Maslov., N.N., "Basic Engineering Geology and Soil Mechanics" Mir publisers, Moscow (1987).
14. M.C., Guire, R K., "Seismic Ground Notion Parameter Relations", Geo-Tech., Division Am., Soc., Civil Engineers, New York (April 1978).
15. Murphy, et al., "The Correlation of peak ground acceleration amplitude with seismic intensity" Bull Seismological Soc., America (June 1977).
16. Narasimham, et al., "Dam failure—A case Analysis Workshop on Environment and Disaster Management", Institution of Engineers, Visakhapatnam (June 1990).
17. Newmark., M.N., et al., "Fundamentals of Earth-Quake Engineering" Prentice-Hall, N J., (1971).
18. Polyakov S.V., "Design of Earth-Quake Resistant Structures", Mir Publishers Moscow (1985).
19. Secretary to Government of India (Statement on Tehri Dam) "The Indian Express" Dated 26-4-1990.
20. Shebalin (quoted from Page 106 of the Text book by polyakov., cited).
21. Tehri Hydro-Development Corporation reports (Vols., I to III) on Environmental Action plans for Tehri Dam., New Delhi (Nov 1989).
22. Trifunac, M.D., et al., "On the correlation of seismic Intensity scales with peaks of Recorded strong Ground motion" Bull., Seis. Soc. Am Feb. 1975).
23. Zeevaert, Leonardo "Foundation Engineering for Difficult Sub-Soil Conditions" Van Nostrand, New York (1973).

8

Impact of Human Activities on Mountain Eco-system

*Mamta Desai**

Abstract

The living organisms in this planet exist under natural conditions in physical and biological isolation. There is a constant interplay both among the different population groups themselves and with the physical and chemical components of the environment. The present paper deals with the increasing human interferences to the nature and ecological degradation in Sikkim Himalayas. In the first part an attempt has been made in some details on the historical background of human settlement in the state, mentioning some aspects of environmental sciences relevant to man and his settlement patterns with the varied ecosystem. In this regard, present trend of population growth and present day landuse pattern have also been discussed in the light of increasing biotic pressure. The second part concerns an analysis of changing attitudes in tackling the vital issues and challenges to human use of natural resources cum

*Professor, Netaji Institute for Asian Studies, Calcutta.

environment in a mountainous region both on mecro and micro scales. In this regard, some changes like human migration etc. have also been discussed to stress the seriousness of the ecological degradation.

Introduction

Sikkim is essentially a Himalayan territory (about 7096 sq km) with the highest and beautiful peaks (6500 m. or more) forms a part of the Eastern Himalayan range. The State is located in between 28°07′48″ and 27°04′46″ north latitude and 88°00′58″ and 88°55′25″ east longitude. Physically it consists of the Greater Himalayan and Inner or Lesser Himalayan zones. The Greater Himalayan region in the north is a snow desert, where highlanders graze their yak and other herds during the summer and lead a nomadic cum pastoralist life. The river Tista and its tributaries deeply dissect the Inner Himalayan Highlands of Sikkim. The river leads and banks of Tista and its tributaries are ferile land. The northern slopes are mainly cold and normally used for pastorage, while the southern slopes are mainly utilised for settlements and agriculture.

The population of Sikkim is composed mainly of Lepcha Bhotias and Nepalis. The first authentic information regarding Sikkim's population in the Census 1891, which shows the then population as 30,458 (Risley—Gazetteer of Sikkim). The population has grown by 1961 to 1,64,000, a phenomenal growth in seventy years.

The tentative figures of 1971 have been shown as 2,09,843. The 1981 Census registers population of about 3,14,999. This rapid growth of population has resulted in development of agriculture, education, industries, towns, settlements and communication system.

Historical Background of Human Settlement

The history of ancient settlement in Sikkim Himalaya still lies in obscurity. The remote past of Sikkim that is known

to-day, comes to us through legends and myths. The Lepcha and Limbu legends do throw light on the earliest chronicle of the state but the same have not yet been authenticated by historians. According to one Lepcha legend, the Rong Pa. (Ravine Folk) or Lepchas, who belong to the Mongolian tribe, were autochthonous to Sikkim. It is necessary to mention that the Lepchas are the indigenous inhabitants, and the people of Tibetan origin called 'Bhotias' took refuge in the country during fifteenth and sixteenth centuries, one of their chieftains was crowned as chogyal or religious and secular ruler in 1642. At that time, Sikkim was quite an extensive country but it lost large areas to Nepal and Bhutan and finally to British. It was the British, who brought Sikkim under the imperial rule.

The earlier human settlements in Sikkim Himalaya were situated along mild mountain slopes particularly facing east and south–east directions and along the Tista Valley. The ideal site was half way up in an airy spur and the main considerations were availability of agricultural land, drinking water and forests for fuel and fodder. As the indigenous Lepchas liked to live close to their lands, their houses in the villages were not very scattered. But the houses were quite open and airy. Each and every house had some open space in the front and in the back. As rainfall is heavy, sometimes metalled sheets were used for roofing along with thatchedroof. The houses were generally of the same pattern, symmetrically arranged in rows and were mainly two storeyed, where the groundfloor was mainly utilised for the animals. All construction materials were locally available, e g. bamboo, slates for roofing etc. The same picture is also found today in the remote corners of the State. Though the houses were not pucca, yet they could withstand the inclement weather and vagaries of nature. Agriculture and animal husbandry were the main stay of the people. So the people were in perfect harmony with nature. Their living was frugal but not precarious like today. Rains seldom failed. There were enough forest lands to meet the fuel and fodder requirements of the people. The population was sparse, and there was a myth that the natural resources were endless.

The situation began to take a turn for the worse during the beginning of the 20th Century. As human population expands in the hills, forests are depleted for the extension of agricultural lands and for the introduction of new settlements and the roadways.

During eighteenth century, the British encouraged Nepali immigration which not only completely altered the population composition of the State but also could not leave the hill eco-system unscathed and became subject to human interferences in a great many ways. When the British left India during 1947, the total population of Sikkim was about 1,50,000 of which half were Nepalis. Within twenty-five years, it became 2,03,000 of which Nepalis constituted 75 per cent of the total population. Now, total population is about 3,14,999 of which 80 per cent is occupied by Nepali community. In this way, the west, south and east districts were filled up by Nepalis by pushing the Lepcha-Bhotia population northward. This large volume of population did not have any alternative other than to depend on agriculture by clearing forest lands. Thus, due to increase of rural population, destruction of forestland started, which has reduced the existing forest cover of the state to a dangerously low level.

Population Growth and Landuse in Sikkim

As said earlier, the population growth in Sikkim reveals that the Lepchas are the indigenous habitants, and the people of Tibetan origin, called Bhotias took refuge in the country during the fifteenth and sixteenth centuries. After British invasion, immigration of Nepalis started. To provide food and shelter to this large population, newer and newer urban centres came up. Table 1 below may explain the growth of population in Sikkim and consequent increase in the number of towns. It appears from Table 1 that prior to 1951 there was no urban centre in Sikkim, and East district has been the only district, with a single urban centre, i.e. Gantak, since 1951. It is interesting to note that the remaining seven urban centres have come up only since 1971.

Table 1

Growth of Population and Towns in Sikkim from 1951-1981

Name of the town	*District*	*Growth of population*			
		1951	*1961*	*1971*	*1981*
1. Mangan	North	—	—	331	780
2. Gantak		2,744	6,848	13,308	36,768
3. Singtam	East	—	—	1,926	4,042
4. Rangpo		—	—	1,785	2,452
5. Namchi	South	—	—	460	1,444
6. Jorethang		—	—	762	3,922
7. Gyalshing	West	—	—	445	750
8. Naya Bazar		—	—	651	962

Source : Census of India : Sikkim Series 1981

The above Table shows that there is a continuous growth in the urban population, particularly in the last decade. The decadal increase of urban population has jumped from 2 in 1951 to 16 in 1981 per 100 persons.

Since full fledged districts in the present form were only formed after 1961, sufficient district wise details are not available. Only those details available at the district level are presented in Table 2.

It appears from the above Table 2, that the North district is the only district with a very high population increase of 1028 persons during the decade 1971-1981. Similarly, its rural population increased to a great extent in the decade 1971-81 as compared to the earlier decade 1961-1971. This

Table 2

District Wise Rural-Urban Decennial Population Increase for Every 1000 Persons in Sikkim (1961-1981)

District	*1961-71*	*1971-81*
North	Total + 334	+1028
	Rural + 300	+1010
	Urban —	+1357
East	Total + 510	+ 613
	Rural + 376	+ 383
	Urban + 1485	+1542
South	Total + 162	+ 423
	Rural + 138	+ 383
	Urban + —	+3391
West	Total —	+ 289
	Rural —	+ 284
	Urban —	+ 553

Source : Census of India : Sikkim Series : 1981.

may be due to the sudden development of animal husbandry and certain commercial agriculture (i.e. cardamom, apple etc.). The East and South districts also have comparatively higher growth rates in rural areas.

The population of the State as per 1981 Census was about 3.16 lakhs which by now must have gone up to more than 4 lakhs. As said earlier, among the districts, the east district is most populated and the north district is sparsely populated. The state has 50,000 holdings of which 37,200

holdings belong to small and marginal farmer representing 64 per cent of the total holdings, but in terms of area it constitutes only 38 per cent of total area. If we examine the settlement report of 1958, it is revealed that the highest percentage of total cultivated land is owned by the Nepalis (66%). The Bhotias and Lepchas own 20 percent and 14 per cent respectively.

Table 3 may give a picture of the existing landuse pattern of the State, which also may give an overall picture in respect of the present status of landuse for various purposes.

Table 3

Present Day Landuse Pattern in Sikkim

Landuse	*Area (000 Ha.)*	*Percentage of total geographical area (%)*
1. Agriculture	98.70	13.53
2. Forest land	265.21	36.36
3. Permanent pastures and culturable wastes	102.49	14.05
4. Land under misc. trees and groves	3.53	0.48
5. Barren and unculturable land	209.01	28.65
6. Land put to non-agricultural use	50.46	6.93
Total	729.40	100.00

Source : Dept. of Agriculture : Govt. of Sikkim.

It appears from Table 3 that only 13.53 per cent of the total area is under agriculture, while forest land occupies a major part i.e. 36.36 per cent. The unculturable waste and barren land (snow-covered mountains, glaciers, lakes, barren hills etc.) occupies 28.65 per cent. The land under non-agricultural use represents common grazing lands and forest land for fuel wood etc. (6.93 per cent).

Crops and Productivity

The area, production and productivty of important crops, grown in Sikkim are given in Tabie 4.

Table 4

Area, Production and Productivity of Different Crops

Crop	*Area (000 Ha)*	*Productivity (Kgs/Ha)*	*Production (000tonnes)*
Rice	19.00	1,211	23.00
Wheat	12.50	1,760	22.00
Maize	41.00	1,415	58.00
Ragi	6.00	1,000	6.00
Barley	1.60	1,375	2.90
Urd	7.00	700	4.00
Rape and Mustard	7.50	827	6.20
Soyabean	5.00	1,400	7.00
Cardamom	22.00	205	4.50
Potato	5.80	5,948	34.50
Ginger	3.00	5,100	15.30
Orange	7.00	2,800	19.60

(These figures represent anticipated achievement by the terminal year of the 7th Plan) (Source : Dept. of Agaiculture, Government of Sikkim).

The principle cash and commercial crops like cardamom, orange, apples constitute an important part in Sikkim's economy. The terrain and climate of this hilly state is suitable for horticulture like all other northern hilly states. Cardamom is grown on wet slopes. Altogether 22000 hectares of land is now under cardamom cultivation of which 7,050 hectares is located in the north district alone. Average yield per hectare is 250 kg, while average annual production is about 4,000 tonnes. Ginger is cultivated mainly in south district. About 5,000 tonnes of ginger is produced annually. North district and some parts of south district are noted for cultivation of apple, orange and pineapple. About 7,000 tonnes of oranges are produced annually. State revenue from horticulture has raised from Rs. 19,560 in 1960 to Rs. 4,43,888 in 1975 without any appreciable extension of planted area. The increase is accounted for mainly phenomenal rise in market price and improved methods of cultivation.

The forest covers about 36 per cent of the total area of the State, and is of immense economic values to the state economy. During the planning periods, efforts have been made to exploit the potentialities as means to increase the state reveuue, but priorities have been given to the better utilization of the Green belt. In Table 5 an attempt has been made to explain the estimated average production of some important iteme of forest produces. The values are on the basis of 1980-81 average prices.

The forest cover of any land is a reflection of its climate. The State and its adjoining areas with varied climatic and ecological conditions in its entire length and breadth support vegetation of different types and species. Except north-western parts, the whole region has a dense forest cover with a rich variety of flora and fauna. The hill slopes are covered with various types of Pines and Sal. The region is also rich in orchids. The northern regions are enriched with oak, birches, rhododendrons etc. It should be mentioned that three-fourths of the rich forest cover have already been depleted.

Table 5

Average Outturn of Forest Products (1980-81)

Item 1	*Quantity 2*	*Rs. Value (000) 3*
Timber	5,720	368
Round Wood	2,700	363
Fire Wood	1,500	526
Charcoal	—	382
Medicinal herbs	—	391
Cardamom	—	117
Others	—	2,100

Environmental Impact Assessment

Due to ever incrcreasing grazing, temperate grass land of the region is diminishing very rapidly. Like other places of Eastern Himalayas, the current position of pastures or grass-lands in Sikkim have resulted from the excessive biotic uses which have brought the potential of grassland production to almost its lowest level. The present level of production of most of the grasslands in Sikkim is about 20 per cent. Generally, a livestock holder has to get a permit from the authority concerned for grazing cattle in hill forests. But unfortunately, they always graze their cattle on steeper slopes, where grazing is prohibited. Usually upto 33 degrees of slope is recommended for agriculture with bench terracing. But land above 50° is cultivated at many places. These have happened, mainly due to the excessive growth of population. Previously, to the Lepchas and Bhotias, agriculture was a subsistence type of economy. Nepalis first introduced intensive terrace farming along with commercial, agriculture. As a result, trade and commerce developed. Moreover, cultivation of

root crops like ginger and cardamom, have certain dangers, as these crops are cultivated along the streams and nallahs, which are wet enough for cultivation, and there is no need for irrigation. As because these are root crops, after harvesting the soil becomes loosened and if there is heavy rain, devastating landslides may occur.

It should be mentioned in this context, that when the British acquired these hilly tracts (Sikkim and Darjeeling), they adopted a policy to maintain the ecological balance in this sensitive Himalayan region. For this purpose, they established most of the tea gardens on the western part of the Tista river, and agriculture was mainly limited to the eastern part of the Tista river. They were also keen to fix the size of the holdings of the agricultural land. Lastly the British made it clear that forest must by preserved for the sake of ecological balance. In fact, for the first time in India, the British introduced a forest policy, in which successful implementation of plantation and commercial crops like tea, apple, orange, cardamom started which resulted in the clearance of large forest tracts, without disturbing the remaining forest cover for ecological reasons. So, inspite of quick development, there was hardly any ecological problem.

As soil erosion is a serious problem in Sikkim, one should be very cautious regarding landuse and felling of trees. In Sikkim, most of the Lepcha and Bhotia houses are made of valuable timber. The report of Bureau of Economics and Statistics confirms that each Lepcha or Bhotia family of five persons requires one tree trunk annually for construction purposes, that means about 70 cu. m. of valuable wood is lodged per house, although less than 20 cu. m. would suffice if properly and efficiently utilised. Moreover in north Sikkim, the Lepcha-Bhotia communities hold livestock in excessive number, sometimes for religious reasons but mainly for commercial purposes. This affects the forest in two ways. Firstly, the forest is used for grazing all the year round, and the cattle feed mainly on the undergrowth or young plants. Secondly, leaves and twigs are collected by the people for the cattle all the year round.

Again, due to scarcity of fuel, especially in rural areas, fuel needs are met by firewood.

Finally a mention must be made of the shifting cultivation, which is still being practised by the Lepchas in north Sikkim. Since, yields on the primitively cleared and cultivated land are low, the area burnt is correspondingly large.

Increasing Mining Activities

The interaction of processes of mineral exploitation with the environment cause ecological imbalance. Due to increasing demand, unsystematic mineral extraction may lead to environmental degradation in the form of land misuse and pollution of air and water. Sikkim is not an exception to such problems.

The polymetallic sulphide deposits of Rangpo area in Sikkim are exposed to Systematic and planned mining. The other polymetallic occurrences at Dikchu in north Sikkim are also to be exploited. The natural geological setting of the mineralized load in Bhotang mines at Rangpo has no problem of overburden to ore ratio as the ore is exposed on the hill scrap with very little or no soil cover (Mehrotra and Joshi 1982).

The geological settings of the region reveal that there are many geological discontinuties in the form of high frequency of foliations and joints, which again is coupled with the incompetent nature of rocks. The problem is really multiplied, when explosives are used for burstings, which many a times cause landslides, debris flow, subsistence and slope failure.

According to Mehrotra and Joshi (1982), there is certain concentration of radioactive mineral in Rangpo ore body. So periodic check of the mine waste and slag is necessary because the concentration of radioactive elements is very injurious to human population. Moreover, the occurrence of Pyrite and Galena have also been reported, which are considered as the

basic source of acid drainage in mining areas. So, it is necessary to monitor regularly the testing of natural subsurface and surface water. The proper dilineation of dumping grounds for the waste is, therefore, important.

Migration

Population pressure in the hill has led to deforestation and erosion, and the situation is becoming more and more critical every year. The lower Himalayan regions are virtually deforested. Migration, in the Himalayan regions is actually a type of cycling; i. e. destruction of forests in the hills, consequent land erosion and search for new lands. These migrations have some direct and indirect effects on the existing human population and their environment. In the past, the transborder trade with Tibet (China) was the main stay of the Sikkimese people like Bhotias and Lepchas. Jhuming was widely practised. Then came the British along with Nepalis. The Nepalis introduced intensive terrace farming with improved agricultural productivity. The growth of population along with increased productivity led to a rise in demand for various consumer goods. As a result industries developed, cities sprang up, and the National Highway removed the Himalayan isolation from the plains.

Man made Hazards

Like all other Himalayan regions, geologically the state is comparatively unstable and young. This unstable geological structure, with steep slopes and high amount of rainfall are the main causative factors which are responsible for the frequent natural calamities in the state. Due to excessive steep slopes, the rivers flood extremely rapidly during times of heavy rainfall. Under such excessive runoff and consequent flooding, erosion progresses at a rapid rate. Extensive removal of vegetative cover due to frequent floods results not only in silt charge, but also in rapid destruction of forest cover. So all the above mentioned factors contribute the erodibility of the hill slopes and a natural weathering process results.

The natural system of weathering is more complicated when man interferes. Actually it is not so with unstable geological conditions or with rainfall, which has been soaking the terrain for millions of years. Any land-form which is not stable enough under the prevailing climatic conditions would have been eroded long back.

The impact of man and technology has been very harmful to the Himalayan ecosystem although it has raised living standards of the people inhabiting the region. This particularly applied to the extraction of forest products like timber, mining and agriculture. So, a balanced development plan is essential, to ensure that the natural environment does not deteriorate any further, and that the rare and spectacular flora and fauna do not die out. Shifting cultivation or jhuming is still practised in north Sikkim, though it is officially banned. With such shifting cultivation, soil fertility rapidly diminishes, and with the loss of protective vegetal cover, massive soll erosion begins. Shifting farmers migrate from place to place destroying valuable forests and as a result, ecological degradation occurs with consequent drastic change from high forest to low vegetal cover.

One thing is very certain now, that the hill people are very conscious, regarding the extension of the plains development concept to the hilly regions. The people of the hill regions and their problems now demand a very scientfic and realistic approach, be it soil conservation, road construction, tourism or technolgical development. All these activities have to take into account the ecological consequences which operate on a vastly different scale from the plains.

References

Census of India, Sikkim Series, 1971, 1981.

Desai Mamata, 1984, Environmental Hazards in Sikkim Himalayas : Himalayan Research and Development, Vol. 1.

Desai Mamata, 1988, Ecosystem and Ethnic Constellation of Sikkim, Best Book Publication Dept. Calcutta.

Lall, J.S. 1981, The Himalaya-Aspects of Change-Oxford University Press.

Misra, R.C. 1985, Environmental Problems created by exploration of mineral resources in Himalaya in Environmental regeneration in Himalaya-Concepts and strategies' edited by J. S. Singh, The Central Himalayan Environmental Association.

Mehrotra, R. C. and M. N. Joshi, 1982, Environmental degradation in Himalayan region due to exploitation of minerals.Nat Seminar on Mineral and Ecology, 15 m (per print).

Shah, S.L. 1986, Planning and Management of Natural and Human Resources in tne Mountains—A micro level approach with special reference to Central Himalaya, Yatan Publications, New Delhi.

Statistical, Handbook of Sikkim, 1981, Bureau of Eonomics and Statistics—Government of Sikkim.

9

Ecological Crisis and Changing Cropping Patterns in India

*Majid Husain**

Cropping patterns and cropping structures are dynamic concepts as they change in space and time. The Indian farmers ordinarily grow different crops in association with each other during the summer and the winter seasons, and change their combination to adjust with the prevailing weather conditions the set objective to optimise their agricultural returns. Every change in the cropping pattern however, has a bearing on the fertility of soils, affecting the ecotopes and ecosystems favourably or adversely.

In recent decades cropping patterns have gone under radical transformation in the various parts of the country. The High Yielding Varieties and the accompanying technology adopted in recent years have resulted in increased agricultural production. Consequently, the production of food-grains have gone up appreciably. The indirect and qualitative effects have been even more significant as the new

***Professor, Jamia Milia Islamia, New Delhi.**

technology has led to noticeable changes in crop pattern, crop structure, mode of production and the use of inputs. The diffusion of High Yielding Varieties has accelerated the growth of market oriented economy, resulting into the destructing of the traditional agriculture. It has not only increased the power of land owners, resulting in the labourers and land-owners polarisation, intensifying class conflict, it has also affected adversely the health of the soils, damaging their nutrients, making the ecosystems more fragile. In many areas the ecosystems are acceleratingly losing their resilience characteristics. The present paper is an attempt to examine the achievements of the new agricultural strategy and to asses the changes that have occurred in the crop combinations. Attempt has also been made to analyse the ecological and socio-eocnomic implications of the emerging crop associations in the various agro-climatic regions.

The paper is based on the data published by the Directorate of Economics, Ministry of Food and Agriculture, Government of India and the field observations made in the various regions by the author. The estimates of area, yield and production for the years 1961-64, 1971-74, and 1981-84 have been used to ascertain the crop combinations of the Pre and Post Green Revolution periods. Moreover, the Economic Survey, and the Annual Reports of the Ministry of Agriculture, Government of India have also been taken into consideration. District has been taken as the unit of study, and in all thirty major crops were considered for computation of crop combination regions. The number of districts, is however, continuously increasing as many new districts have been carved out over the period of the last twenty-five years. For a comparision of data of acreage and yield of crops, adjustment in the reported data has been made with the Pre-Green Revolution period when the new districts were not existing.

Owing to its large latitudinal extent and altitudinal variations, India has great diversity in rainfall, soils, geo-ecological and socio-economic conditions. These variations have collectively affected the crop concentration and crop associations resulting into an intricate cropping mosaic which demand definition and geographical explanation.

Achievements of the New Agricultural Strategy

As stated at the outset, Indian agriculture and rural scenario have changed substantially after the adoption of new agricultural strategy. The major achievement of the new strategy is the boost in the production of major creals e.g. wheat and rice. This fact may be testified from the statistics given in Table I.

Table 1

Indian Growth Pattern of Foodgrains (million tonnes)

Crop	*1960-61*	*1970-71*	*1980-81*	*1987-88*
Rice	35	42	54	60
Wheat	11	24	36	46
Total cereals	69	96	119	138
Pulses	13	12	11	13
Total foodgrains	82	108	130	157

Source : Government of India, Ministry of Agriculture (Annual) Report, (1987-88).

An examination of Table I vividly shows the increase in rice production from 35 million tonnes in 1960-61 to 42 million tonnes in 1970-71 which increased to 54 and 60 million tonnes in 1981 and 1987 respectively, signifying an incredible achieve ment. The production of wheat which was only 11 million tonnes in 1960-61 rose to about 38 million tonnes in 1980-81 and went upto 46 million tonnes in 1987-88, registering an increase of over 300 per cent in the total wheat production. Part of this increase may be attributed to an extension in wheat acreage, but the yield of wheat rose from 500 kgs in 1961 to about 1700 kgs per hectare in 1987-88 signifying an increase over 300 per cent. In other words wheat responded exceptionally well in the High Yielding Varieties Programme, while rice was below the average of all food grains. The growth rates were negative in pulses, gram, cotton and jute.

It may also be noted from Table I that the production of pulses has been stagnant being only 13 million tonnes in 1987-88 as against the same tonnage in 1960-61. Consequently, the per capita availability of pulses has fallen from 78 gm per head per day in fifties to only 36 gm today.

The Percentage distribution of cereals output has been given in Table II. An examination of Table II reveals two significant changes in the production of cereals. Firstly, the output of cereals has risen at the rate of 3 to 4 per cent per annum but the output of pulses has remained stagnant or even declined. This has resulted in a decline in the importance of pulses in foodgrains output from 16 per cent in 1960-61 to only 9 per cent in 1987-88.

Table 2

Percentage Distribution of Cereals Output

Year	*Rice*	*Wheat*	*Coarse grains*	*Total*
1960-61	50	14	36	100
1970-71	50	16	34	100
1980-81	44	25	31	100
1987-88	44	32	24	100

Source **:** ***Economic Sruvey*** (Various issues).

As against this, cereals have risen in importance from 85 per cent to 91 per cent during the same period. Secondly, among the cereals there is a decline in the proportion of rice from 50 per cent to 44 per cent during 1961 to 1987. But during the same period the proportion of wheat has more than doubled i.e. 14 per cent to 32 per cent. Another noteworthy change is the decline in the share of coarse grains from 36 per cent in 1960-61 to 24 per cent in 1987-88. The growing pattern of wheat indicates a substitution of coarse grains with wheat, both on the side of production and consumption.

Changing Patterns of Crop Combinations

The combination of crops and crop diversification are the useful indicators of the degree of diversification. Excepting the areas of very heavy or very scantly rainfall, the large number of crops entering in the combination indicate ordinarily a subsistence agriculture, while the less number of crops denote the market oriented economy of the rural populace. Thus for a comprehensive understanding of the agricultural mosaic of an agro-climatic region the study of crop combination is of great significance.

In recent years the concept of crop combination has increasingly engaged the attention of geographers and economists. The studies made about crop association range in approach from topical to regional and very in extent from small areas of micro political units to the entire country. The different methods applied in the delineation of crop combination regions can be summed up under two headings. The first one is the arbitrary choice method, e.g. the first crop only, the first two crops only, or the first three crops, etc. The crop association regions delineated on arbitrary choice method are, however, not rational and judicious as the remaining crops are irrationally excluded without consideration of their percentage weightage in the total cropped area. The second method is statistical, developed in terms of variables based on certain differences which are relative and not absolute. This method being based on statistical approach is more accurate, reliable and scienctific as it gives better objective gronping of crops of the given region.

In the present paper for the delineation of crop association regions of India, the standard deviation technique has been applied which may be expressed algebrically as under :

$$SD=\sqrt{\frac{d^2}{n}}$$

When d is the difference between the actual crop percentages in a given areal unit and the appropriate percentage in the theoretical curve and n is the number of crops in a given combination.

Where the deviation from the actual percentage is lowest that is taken as the cut point for the crops which enter in the combination. The results thus obtained have been plotted in the maps of crop associations for the years of 1961-64, 1971-74 and 1981-84. A comparision of these maps vividly reveals the changes that have occurred in the crop combinations, indicating the regions and districts in which the agriculture has moved from subsistence farming to market oriented economy.

Crop Combination

Monoculture

Single crop dominance or monoculture in the Pre-Green Revolution period (1961-64) was found in thirty four districts. The monoculture region of rice was sprawled mainly in Assam, Andhra Pradesh, Bihar, Orissa, Tamil Nadu and West Bengal. Although rice is grown under the varried physical conditions, yet it thrives well in the hot and humid areas occupying fertile alluvial soils. Bajra was a monocrop in the districts of Barmer and Jaisalmer, both lying in the Thar-desert. In these districts the average annual rainfall is around 20 cm with a very high degree of variability of rainfall. Under such a harsh and arid environment only the less moisture requiring crops like *Bajra* (Bulrush-millet) perform well. Cotton had the main concentration in the district of Sholapur of Maharashtra which has the predominance of black earth soil.

In 1971-74 the cropping scenario got changed substantially as monoculture covered 88 districts. Rice, bajra and cotton remained the monoculture crops of the country. Rice as a monocrop occupied 78 districts in 1971-74 as against 32 districts in 1961-64. The area of Bajra as monocrop increased in the districts of Barmer, Bikaner, Churu, Jaisalmer, Jodhpur, Nagaur and Sikkar (Rajasthan) in which it occupied over 70 per cent of the cropped area. Cotton emerged as the single most important crop in Jalgaon and Sholapur districts (Maharashtra).

In 1981-84, the number of monocrop districts increased to 90 from 81 in 1971-74. The nine new districts added in this list emerged as the rice monoculture areas. These districts lie in Assam, Bihar, Tamil Nadu and West Bengal. The position of cotton and Bajra growing districts, however, remained unchanged. The increase in the rural strength of rice reveals that the High Yielding Varieties of this crop performed appreciably well which attacted the farmers to bring more area under its cultivation.

Two Crop Combination

Two crop combination aprawled over 89 districts in 1961-64. Rice, wheat, maize, pulses, bajra, cotton, sugarcane, tea, coffee and mustard combined in different associations to form two crop combination. Rice with pulses (gram/lentil) was the single most important two crop combination which covered fifteen districts, namely, Balaghat, Bastar, Bilaspur, Gaya, Kanyakumari, Krishna, Malda, Murshidabad, Muzaffarpur, Patna, Raigarh, Raipur, Ranchi, Shahabad and Santhal-Pargana. Maize wheat, wheat-rice, Bajra-pules, wheat-sugarcane, wheat-jowar, and rice-maize are the other sub-groups of two crop combination in the pre-Green Revolution period. In the post-Green Revolution period as many as 121 districts entered in the two crop combination, an addition of forty districts over the 1961-64 areal units. This combination is traceable in all the states of the country for which the agricultural data was available. The constituent crops of this combination however, vary from one agro-climatic region to another, owing to the prevailing temperature and moisture conditions. In 1971-74, wheat and rice emerged as the dominant crop association, occupying several districts of the Great Plains of India.

In 198I-84 the number of districts which emerged as the two-crop combination areal units rose to 156, registering an addition of 66 districts within a period of ten years. This tremendous increase in the number of districts having two crop combination in the post-Green Revolution period vividly shows that the farmers, especially that of the Gangetic Plain are increasingly concentrating on less number of crops and thus they are moving towards the commercialised agriculture. In fact, in most of the districts of Punjab, Haryana, and Western Uttar Pradesh rice and wheat dominate the *Kharif* (summer) and *Rabi* (winter) seasons respectively. Both these crops are however, soil exhaustive which is detrimental to the health of the soils, affecting the health of the soils adversely.

Three Crop Combination

In all 23 districts entered in the three crop combination in 1961-64, the constituent crops being rice, Jowar, wheat, *Bajra*, maize, pulses, sugarcane, cotton and groundnut. Wheat, Jowar-gram was the major combination sprawling over the districts of Hamirpur, Indore, Jhabua, Jhansi and Kota. Rice-wheat-gram association was significant in Panna and Satna districts of Madhya Pradesh. There were 15 more districts entering in the three crop combination but the constituent crops in each district varied from each other.

In 1871-74 three crop combination was found in 32 districts, the constituent crops being more or less the same. In 198 -84 the number of areal units in three crop combination increased to 51, registering an increase of 19 districts in ten years time. The state of Uttar Pradesh, Haryana, Bihar and Madhya Pradesh shared most of the districts falling under the three crops association units. Before the diffusion of the new seeds in many of these districts four to seven crops were constituting the combination, but with the popularisation of new agricultural technology, the farmers concentrated on wheat and rice and the less important grains, coarse cereals and pulses were discarded with by them.

Four Crop Combination

Four crop combination was found in 31 districts in 1961-64. In this combination 19 crops grouped to make different combinations. Wheat-rice-gram-pulses, occupying the districts of Barabanki, Mirzapur (U.P.), and Jabalpur and

Table 3

India : Comparative Picture of the Various Crop Combination Areal Units

Combination of crops	*Number of Districts*		
	1961-64	*1971-74*	*1981-84*
Monoculture	34	81	90
2-Crop Combination	89	121	156
3-Crop Combination	23	32	51
4-Crop Combination	31	25	21
5-Crop Combination	28	24	22
6-Crop Combination	30	09	09
7-Crop Combination	23	05	—
8-Crop Combination	25	02	—
9-Crop Combination	11	—	—
10-Crop Combination	06	—	—

Source : Computed from the agricultural statistics by the author.

Rewa (M.P.) can be said as the major four crop combinatino of the pre-Green Revolution period. In 1971-74 the number of districts assigned to four crop association got reduced to 25, which further declined to only 21 districts in 1981-84. It is however, interesting to note that throughout the period of study the four crop combination is conspicuous in the areas of high variability of rainfall. The farmers of scantly rainfall areas grow several crops in the hope that either of the sown crops will be a good one.

Five Crop Combination

In all twenty-eight districts were assigned as the five crop combination areal units in 1961-64, the constituent crops being rice, wheat, sugarcane, maize, gram, Bajra, pulses, cotton, groundnut, jute, mustard and rapeseed. Rice-wheat-sugarcane-maize-gram was most significant in this combination occuying the districts of Kheri, Muzaffarnagar Pilibhit, Rampur and Saharanpur. All these districts are situated in the north-western parts of Uttar Pradesh to the south of Siwaliks. The remaining 23 districts entering in this combination are well scattered in the various parts of the country.

In the post-Green Revolution period (1971-74) only 24 districts were designated under five crop association which got further reduced to 22 in 1981-84. In most of the districts where five crop association is found, assured means of irrigation are not available and the rainfall is scantly as well as erratic in nature.

Six Crop Combination

In 1961-64, thirty districts got the designation of six crop combination. In all 21 crops formed diversified groupings of this category of combination. Wheat-gram-barley, maize-millets-pulses were the major crops which entered in the association in Madhya Pradesh, and eastern Rajasthan. In other parts of the country this combination got modified as new crops replaced some of the crops of the above given association. After a decade in 1971-74 only nine districts were designated in this combination, and the picture remained unchanged in 1981-84 when the same number of areal units remained in this category, though the districts were not the same. From thirty areal units to nine districts is a significant decrease which strengthens the hypothesis that the Indian farmers are doing away with the highly diversified cropping patterns.

Seven Crop Combination

Before the adoption of High Yielding Varieties in 1961-64 as many as 23 districts were assigned in seven crop combination, which got reduced to only five districts in 1971-74. It is interesting to note that this combination disappeared from the agricultural mosaic of India. At present, there is not even a single areal unit in which seven or more than seven crops constitute crop association, thereby indicating a high degree of market oriented economy of the rural populace.

Eight, Nine and Ten Crop Combinations

Eight crop combination occupied twenty five districts , 1961-64. A significant feature of this association is that in all the units, combination of crops differ from each other. Two districts, namely, Hoshingabad and Morena (M.P.) retained this combination, the constituent crops being wheat, jowar, cotton, groundnut, maize, barley, rice and *Arhar* (pulse).

The nine and ten crop combination which were occupying eleven and six districts respectively in 1961-64 disappeared from the cropping structures. In brief, by applying the standard deviation teehnique on the agricultural data available at the decadal interval shows that at present out of the 360 reporting districts only 59 districts have four to six crop combination as against 154 out of 310 districts in 1961-64. The complete elimination of seven to ten crop combination establishes beyond doubt that the Indian farmers are increasingly interested in few number of crops and they are progressively moving towards the commercial agriculture. As a matter of fact, the agriculturally more advanced states of Punjab, Haryana and Western Uttar Pradesh are having heavy concentration of wheat and rice.

Socio-Economic Implications of Changes in Crop Combinations

Although changes in cropping patterns and reduction in the crop combinations have boosted the production of major cereals (wheat, rice), yet they have not benefitted the large and small farmers and landless labourers alike. As a result of the adoption of the new agricultural strategy and changes in the cropping patterns had adverse effects on the distribution of income in rural areas. Thus they have contributed to widening the disparities in income between different regions, between small and large farm, and between land owners on the one land and landless labourers and tenants on the other. Only the small minority of cultivators with holdings of ten acres or more have been in position to mobilise surplus capital for investment in land development. Moreover, this class has pyramided its gains by using increase profits by .... to buy more land, improve land already under cultivation, and purchase modern equipments. The big farmers have multiplied their assets. This tremendous increase in agricultural assets of a small group of farmers in each village, especially in those situated in the more productive plains has resulted into great social and political tension which is breaking the traditional mode of life of the Indian rural populace and removing the good social institutions like mutual co-operation and help at the time of crisis illness and acutelabour demand. The importance of land has increased considerably, and scarcity of land has been increasingly felt in almost all the agricultural tracts. The small and marginal farmers have disposed of their lands and migrated to the neighbouring urban places to start a new life. It led to social tension and conflicts between haves and have nots. The incidence of rural violence has increased which is the result of the widened income chasm between landlords and landless workers. Violence has extended from the field of politics to the depth of society—agrarian and caste relations, inter-communal relations, and family affairs. These violences are the expressions of economic deprivation of poor people of rural India which have been accentuated owing to egregiously unequal distribution of income after the adoption of new agricultural technology and changes in the cropping patterns.

Ecological Implications

The traditional cropping patterns evolved on the basis of empirical experience by the Indian farmers has almost been completely transformed. The rotation of soil exhaustive crops following by the soil enriching crops has been given up. For example, over the greaterparts of Punjab, Haryana, Uttar Pradesh and Madhya Pradesh, millets and maize (summer crops) used to be followed by gram (winter crop), the previous being soil exhaustive and the later soil enriching crop; or land used to be followed before wheat was sown. Now after the installation of tube-wells and development of canal irrigation farmers are largely devoting their arable lands to rice in *Kharif* (summer) season and to wheat in *Rabi* (winter) season. Both these crops are soil exhaustive, affecting the fertility adversely. The States of Punjab, Haryana and Western Uttar Pradesh where the average annual rainfall is below 80 cm have emerged as the leading producers of superior quality of rice, locally known as *Basmati* —which requires around 150 cm of rainfall. In order to optimise their agricultural profits, the farmers are growing two to three soil exhausting crops in one year in the same field. Such a practice is depleting the natural fertility of the land at an accelerating rate. Consequently, the land has become hungry. There is a general complaint of the farmers of Punjab, Haryana and Uttar Pradesh that each year they have to apply progressively more chemical fertilisers, insecticides, pesticides and irrigation water to obtain good crops. This is clear testimony of the fact that natural fertility of arable land is declining at a faster rate, which is a cause of conern for the farmers and planners alike.

The market oriented agriculture in which wheat, rice, and maize or sugarcane are prefered by the growers, demands assured supply of irrigation, especially in the areas of deficient rainfall like that of North-western parts of India. Canal and tube-well irrigation have been extended in such areas. For example, the Indira Gandhi Canal is irrigating the Ganganagar, and Bikaner districts of Rajasthan desert. Irrigation works have also been extensively developed in the semi-arid districts of Haryana i.e. Sirsa, Hissar, Bhiwani and Mahendergarh. The farmers out of need and excitement over-irrigate their crops. Consequently, many good tracts of fertile land have become affected adversely by the saline and alkaline formations. Such tracts are quite numerous in Ludhiyana, Jallundhar, Patiala, Faridkot, Amritsar, Firozpur districts (Punjab), Hissar, Bhiwani, Sirsa, Gurgaon, Panipat, Kurukshetra, Sonepat (Haryana), Meerut, Bulandshahar, Aligarh, Etah, Etawah, Mainpuri, Farrukhabad, Kanpur, Fatehpur, Allahabad, Hardoi, Partapgarh, Moradabad, Shahjahanpur, Budaun, Lucknow, Barabanki, Rampur and Bareilly etc. in Uttar Pradesh. The irrigated districts of the Thar-desert i.e. Ganganagar, Bikaner, Jaisalmer, Churu, have developed numerous tracts and patches of saline and alkaline affected soils. Water-logging is a serious problem in newely irrigated areas of Rajasthan desert. Similarly, the adverse effects of changes in traditional cropping patterns in the states of Bihar, West Bengal, Madhya Pradesh, Gujarat and Maharastra are quite conspicuous. The State of Tamil Nadu and coastal districts of Andhra Pradesh are not an exception to the fast declining fertility of soils. Some of the adversely affected areas, as a result of change in cropping patterns and crop association have been plotted in Fig. 4. This Figure clearly shows that in certain areas over irrigation is leading to waterlogging, while in other areas the chemical properties are getting transformed resulting into saline and alkaline formations. The over application of chemical fertilisers is destroying the micro-organism of the soils which are so vital for maintaining the fertility of the soils. Succinctly, the changes in the cropping patterns though giving more agricultural returns, is injurious for the health of the soils, creating ecologically adverse conditions which may render the fertile plains, unproductive, infertile and barren, if remedial steps are not taken.

References

Husain, M., 1982, *Crop Combinations in India,* Concept Pub., Co., New Delhi.

Raj, K.N., 1984, *'The Green Revolution'.*

Rao, V.K.R.V., 1974, *New Challenges before Indian Agriculture,* Panse Mimorial lecture, April 1974 (Mimeographed).

Hanumantha Rao, C.H., 1975, *Technological Change and Distribution of Gains in Indian Agriculture.*

Singh, D.P., 1980, 'The Impact of Green Revolution on the Poor' *Agricultural Situation in India,* August, 1980.

Minhas, B.S., T.N. Srinivasan : 'New agricultural Strategy Analysed' *Yojan,* Jan. 1966.

Chandan, A., 'Victims of Green Revolution' *Economic and Political Weekly,* 23, June 1979.

Saini, G.R., 'Green Revolution and Distribution of Farm Incomes, *Economic and Political Weekly,* March 1976.

10

Industrial Pollution : A Case for Developing Countries

*Dr. Clement M.P. Oniango**

Introduction

As the quest for modernization through industrialisation reaches the climax, industrial pollution which is a sad state of environmental degradation becomes a concommitant reality of the said process. This process is what the author of this article with refer to as the *inappropriate paradigm of development with destruction*—a paradigm not geared towards a holistic, sustainable process of development by the population of developing countries. In this paper, it will be argued that populations of developing countries *must* develop a new and appropriate paradigm of development *without* destruction (industrialise without industrial pollution). This is a paradigm geared towards a holistic, sustainable, self-reliant, equitable and participatory process of development for all developing countries in the world for the survival of their species on earth. But one might ask : how can this be achieved ?

---

*Dean, Faculty of Social Science, University of Nairobi, Kenya.

The first step is that while we are industrialising now, we must look at ourselves now and not beyond the present generations. Second, we must learn to look within our national boundaries as we seek for meaningful sustainable industries of development. At the same time, we must keep cognisance of the dire need for feasible solutions to the pressing global environmental concerns of today. We shall call this collective responsibility is shaping our common destini. Much of this collective responsibility will hinge upon our success in radically transforming the existing consumption patterns, life-styles, production as well as distribution schemes, individual actions and attitudes, as well as changing our present socio-economic and political structures of power and decision making.

While we are acting locally in our development plants of industries we must ivoke the maxim of thinking globally. Why ? Because while we struggle with environmental problems, we must remember that the said problems respect no national or geographical barriers. The unavoidable dialemma that developing countries must grapple with is that the process of industrialization has built in mechanism rendering the population of these countries "endangered species." If they do not industrialize, they starve but if they industrialize *without* care, they suffocate themselves with industrial pollution. The only solution to this problem is to industrialize *with care*. The existence and destiny of the endangered species of the developing countries will be charted by our own individual and collective actions as well as inactions. In other words our future lies in our own hands and in particular it will be dictated by the seriousness of purpose we shall put our participatory sustainable development by industrialising without destructions.

Factors for Implimentation

The Temporal Factor

It is quite tempting for the members of developing countries to feel that we are behind in pollution by same degree that we are lagging behind the developed world in terms of industrialization. This may be so in relative terms but does not suggest that we do not suffer from industrial pollution.

It would be bad reasoning to argue that because the developing countries were late in industrialising, they can, therefore, take their sweet time in grappling for solutions to the pressing global environmental concerns of today. In this context, there is simply no point in procrastination with regard to solution to the problem under references—the time to act is now or else we run the risk of self-extinction.

The Right to Life Factor

While it is true that the right to life is the basic right from which all other rights emanate, our duty of guarding and guaranteeing this fundamental right to life for ourselves as well as for the future generations must be seen as our paramount responsibility then we do not deserve the right to life. We must win this right to life through our demonstrated responsibility as custodians or else we forfeit the right to life. This responsibility calls for care and all the life-support systems.

The Numerical Factor

It is noteworthy that the most "endangered species on earth" ... population of the developing countries constitutes more than three fifth of the entire human race ... more than three billion people.[1] This so-called *Third World* should not be taken for granted since a numerical factor of this magnitude, will of necessity, (in the absence of care) have corresponding pressure of demand on earth, such as, the pressure of fulfilment of basic needs. This pressure will continue to build up as these numbers swell, thereby increasing environmental strain and tension. While, in response to these pressures, various industrial kinds of "development" are coming up, millions of destitute people in the developing countries to live in abject poverty. In the name of "progress of development", industrial pollution in the developing countries worsen the spiral of poverty and impoverishment afflicting the millions of the numbers of the developing countries.

According to the UNEP information[3], the socio-economic and ecological ills, inequalities and imbalances that plague the earth is staggering, both in frequency and in severity both in industrialised and Third World countries as tabulated hereunder :

—Asia has over 50% of the world's farming families, but only 70% of the world's farming families, but only 25% of the world's arable land and over 65% of the world's poor and hungry.

FAO estimates that without effective long-term conservation measures, soil erosion threatens some 544 million hectares of cropland in Third world countries.

In 1988—more than eight years into the UN-initiated "International Water Decade"—nearly 50% of the population in Third world countries were still drinking contaminated water.

With only 25% of the world's population, industrialised countries consume 80% of its natural resources and generated 75% of the estimated 2.5 million trillion metric tons of municipal and industrial wasted produced worldwide annually.

Third world countries account for only 20% of global pesticide use, but account for 50% of the world's pesticide poisonings and 90% of the world's pesticide-related death.

The Third World's external debt crisis is currently estimated at US $ 1.3 million. By the year 2010, Africa's external debt is estimated to be more than 28% of its total export income. This explains why the developing countries are in dire need and hurry to industrialise at any cost, even at the expense of industrial pollution.

Of an estimated total area of 2 billion hectares of tropical rain forest worldwide (this is what is known as The Third World (Belt), some 11 to 15 million hectares are estimated to be lost each year—an area equivalent to Austria. In other an area of tropical rain forest equivalent to 20 football pitches is lost from developing countries every single minute. Global deforestation occurs in Brazil, Colombia, Peru, Mexico, Venezuela, Ecuador, Ivory Coast, Nigeria, Zaire, Indonesia, Thailand, Malaysia, India and the Philippines. It is not by coincidence that these 14 countries command 50% of the Third World debt. The charcoal industry in developing countries worsens this situation.

Due to deforestation, 20% of the five to ten million species known to science now are likely to be extinct by the year 2000. If the present trend of species extinction is not curbed, more than 50% of species will disappear in another 50 years. It is in this desperate understanding that the Kenya Government fighting for the survival of special species of our fauna, such as, the *rhino*, *cheetah*, *elephant*, etc. The poaching activities are a menace to this state of affairs.

Given that on the average, each person in an industrialised country consumes eight times as much energy as a person living in the sub-Sahara Africa, the African will be worse of in the near future when the source of energy in Africa will diminish (e.g. fire wood) and become too expensive to be afforded by an ordinary African in the sub-Sahara. Essentially, the constituents of the sub-Sahara Africa will continue to be poorer and pooper with time unless the deteriorating ecosystem is improved.

Industrial pollution affecting natural resources like land, water and forests in developing countries will be a *serious threat* to the constituents of the Third World countries. Why ? Because the poorest of the poor in the Third World countries are most dependent directly on natural resources, such as, land water and forests. As stated above, it is this same resource base that is witnessing rapid deforestation, water pollution, soil erosion and extinction of natural species whose extinction will result in inbalance of ecosystem. In addition to all these deteriorations, the Third World countries have geometric population increases which put stress on proportionally decreasing acreage of land which is simultaneously deteriorating.

In pursuit of foreign currency, the developing countries have sought for and received financing for large-scale projects, such as the Mumias Sugar Company,[3] the Webuye Paper Mill to name but a few.

The artificial chemical fertilizers, and pesticides in the sugar cane belt in the Mumias area have far-reaching polluting effect on the drinking water, The Webuye Paper Mill in Bungoma has and abominable stench covering 3 to 5 km from the plant; All this is industrial pollution going on without conservation measures. While these large-scale projects are appropriate to the needs of alleviating the standard of living of the Kenyans as well as earning foreign currency as cash crop industries, they pose serious industrial pollution problems which call for arrest.

Other areas of concern that are aggravating environmental degradation are : Slaughter houses which have open trenches leaving blood to flow, again, with abominable strench; Saw mills whose saw dust is dangerous for safe breathing air, open sewage leading to rivers, such as the Nairobi River in the city of Nairobi.

APPROACHES TO SOLUTION

The Education Approach

(Environmental Awareness)

People in the developing countries of the world in Africa, Asia and Latin America should be made aware that millions of poor people in these countries go hungry not necessarily because the world lacks food, but because they lack the means to grow or purchase it. An obsession with industrialization is not the answer to our problem. We need to address ourselves to the factors contributing to their problem, such as, landlessness, Third World foreign debt from industrialised countries.

We need to alert our policy makers that several options could be explored to mitigte our problem of development. For example: better terms of trade between industrialised countries and the developing countries. These terms should be aimed at encouraging farmers in developing countries to grow more food and especially traditional food to satisfy the needs of their own countries. This should be the spelled out priority while the export of cash crop is maintained as secondary.

The reallocation of land to the landless poor with priority for food crops for domestic consumption should also be emphasized. There should also be an effort to an end the burden of foreign debt, and a curb of the drain of natural and other resources from developing countries to industrialised countries.[1]

It is common knowledge that the Booker London running the Mumias Sugar Industry in Kenya want their sugar rapidly and cheaply. Their business shares their interest in maximum quick returns. But the rural poor who are indigenous Kenyans who constitute the bulk of the population who sacrifice their land will bear the long-term costs arising from this industry such as, exhaustion of the natural land fertility due to strong chemical fertilizers, loss of their traditional pattern of agriculture of food crops which is now replaced by luxurious cash crops, pollution of their drinking water due to chemical fertilizers, extended periods of hunger in the year since it takes minimum of 18 months for the cane to be ready and another six months for harvesting.

Participatory Action and Accountability

All groups of people must be seen to participate in the protection of the biosphere : air, water, and earth; we must minimize and safely dispose hazardous wastes common people should dutifully reject their back-yards being used as disposal areas of hazardous wastes by industrialised countries even when their leaders have received kickbacks. Where this has happend "The People" should demand for damage compensation. We should all carry the responsibility of managing our environment.

Ethical Concern For the Environment

Developing countries should take cognisance of ethical considerations for the environment when formulating their national policies for development. They should make sure they are getting a fair deal for their poor nationals when making industrial deals with developed nations, making sure that the following are satisfied :

: that in quest for industrialization, the health of their nationals is not endangered.

: that these industries do not cause damage to the environment during manufacture, use or disposal.

: that they avoid for their industries, materials derived from threatened species or from threatened environments e.g. ivory industries, skin industries etc.

: that they avoid promoting industries dependent on use or generation of toxic chemicals and hazardous substances,

: that they do not hide information to communities concerned about the potential hazards associated with high-risk industries for instance, bodies such as the Kenya Bureau of Standards maintain their integrity in checking and enforcing regulations relating to health and safety of consumers as well as quality of the environment.

Implementation

There is dire need of implimenting all the above suggestions through a fostering and institutionalising body. For the time being, a department or Institute at the University of Nairobi would suffice.

In the Department of Philosophy and Religious Studies, University of Nairobi, we teach Environmental Ethics apart from other Applied Philosophy subjects like Medical Ethics, Business Ethics, etc. We would therefore, like to link with Centre for Environmental Management and Planning with a view of fostering research, dissemination of knowledge on environmental issues to the public as well as Policy Makers, organising symposia on these issues and above all participating in activities nation wide regarding Environmental Management and Planning. This suggestion will, if accepted and endorsed, will go a long way in implimenting the ideals Environmental Management and Planning in Kenya which, like many developing countries *must* industrialise. We want Kenya to industrialise without destroying itself. It is only through such an Institute that Kenya will be guided to avoid the mistakes the North made in their process of industrialization. Kenya needs a *New Vision*; a vision to industrialise with a *conscience*. This Institute will work hand-in-hand with Kenya government which is against bribery and corruption.

The Institute will, for example, formulate theoretical and practical models around questions, such as :

(a) "Are our industries ethical ?"

(d) "Do the manufacturing companies engage in business practices which take advantage of lenient legal, social and environmental regulations in our host country as compared to the strict regulations existing in the companies' home country ?"

(c) Do the companies engage in malpractives like bribery and corruption ?"[5]

(b) "Is the production ecological does the production process care for the environment ?"

(e) Do the business enterprises qualify to be judged for their social responsibility by the following criteria;

(i) that they respect the environment ?

(ii) that they are mindful of and sensitive to the Kenya cultural diversity ? If it is a cash crop industry will it undermine the traditional food crops ?

(iii) that they will be accountable throuph disclosure of information ?

(iv) that they will be relevant to the needs of the local community ?

There have been cases of large-scale projects which have proved inappropriate to both the needs and the local conditions of developing countries...thus aggrevating environmental degradation, natural resource depletion and socio-econmic disparity.

(v) that they will enforce employment policies and fairness to workers including health and safety standards in the workplace.

(vi) We shall propagate a prescriptive philosophy alerting the policy-makers to make laws to guide business enterprises to conduct themselves socially and ecologically responsible in producing and marketing their products.

(vii) We shall high-light for the full implementation of the United Nations "Guidelines on Consumer Protection" and adoption of the "UN Code on Transnational Corporations."

Conclusions

In this paper an attempt has been made to show that industrial Pollution is a reality especially in any society where man is constantly searching hungrily for new profits. It has happened in the industrialised North. But we *must* minimise it in the industrialising South. Unless we take up this responsibility, we are threatened of extinction. We must act now if we expect to survive tomorrow. All of us must participate in this exercise. I have proposed that we institutionalise this process by establishing the Institute of Environmental Studies and Management at the University of Nairobi to be housed initially in the Department of Philosophy and Religious studies where we currently teach Environmental Ethics. This Institute, it is observed, will have objectives of Research, Action, Policy and Participatory orientation for the welfare of the earth, man, and constructive industrialization.

References

1. The New Universal Library, Volume III, International Learning Systems Corporation Ltd., London, 1969.
2. See UNEP/WHO *Agricultural Residues; Environmental Pollution Control*, WHO Offset Publication 30, ISBN 92 4 1700300 65 pp 1976.
3. See Oniang'O C.M.P. (et al) *The Socio-Economic Profile of the Mumias Sugar Scheme*, 1987, Nairobi, Kenya.

4. See H. Knop (ed) *Environmental Management* UNEP/II UNEP/II ASA (1975) and also V. Ranganathan (1977) *Environmental Policies and their Implication for Trade and Development : A Case for India*, UNCTAD FP/0404—78—03. Both authors discuss definitions of Environmental Management, the strategy of environmental management in relation to the countries affected : criteria for assessment and evaluation of impact.
5. Daniel T. arap Moi, *Kenya African Nationalism Nyayo Philosophy and Principles*, Macmillan Publishers, 1986.

11

Environment and Sustainable Development

*Radhamohan**

It seems to be a fraternal conflict. The words ecology and economics although owe their origin to the same Greek word 'OİKOs', i.e. habitat or household are at loggerheads for a long time. Contradictions between the two arose due primarily to the market orientation and myopic considerations. 'Economic development appeared to go hand in hand with ecological destruction'.[1]

Hence for a long time economic development and conservation of environment were considered incompatible. Only recently it is realised more and more that environment and development are only two sides of the same coin and development efforts cannot be sustained without a deep concern for conservation of natural resources. It is also increasingly acknowledged that humans constitute only a part of the great web of life and the well-being of the human beings and well-

* Senior Scientist, Department of Science and Technology, Govt. of Orissa, Bhubaneswar.

being of non-human being life is closely interlinked.[2, 3] By and large enlightened opinion converges today on the fact that an environment unsuitable for birds and animals would also be unsuitable for human.[4]

For this convergence of view to take place, we took rather a long and winding road. To take environment first, initially it was John Muir, the father of the USA environment movement who expressed his concern about the vanishing wilderness of the USA. To Muir Christians impeded a proper appreciation of natural world for they had dispensed with nature Gods in favour of a single diety in whose images we were made. It was also 'stingy' in the way, it saw only human is having immortal soul and a view that enabled the manipulation and domination of the non-human world for human ends.[5]

Later in sixties Rachel Carson in her book 'Silent Spring' melodramatically exposed the dangers of reckless use of pesticides and insecticides and that triggered a great deal of concern about protection of our environment. The book also changed our view that we could remain safe while harming and eliminating others. More and more people believed that we should set aside our attitude to 'conquer' nature and that we were simply 'plain members of the biotic community, fellow voygers, with other creatures in the odyssey of evolution' to use the words of Aldo Leopold.

Environment movement gathered strength after the first International Conference on Environment held in Stockholm, Sweden in the year 1972. A series of international conferences thereafter, sustained public awareness compaigns, formation of UNEP, educational campaigns by number of international NGOs, highly published books like 'Limits to Growth', 'Small is Beautiful', 'Only One Earth', publication of a large body of journals, research papers, organisation of a number of seminars, workshops, symposia etc. and above all a series of environmental disasters like Minamata disaster of Japan, Bhopal tragedy, reservoir-induced earth-quakes, nuclear accidents like Chernobyl of Soviet Union further fuelled the environment movement.[6]

World-wide concerns about 'green house' effect, ozone layer depletion, acid rain, alarming decline of tropical rain forests and erosion of genetic diversity, desertification, air and water pollution, urban blight provided further impetus to the environment movement and today suddenly we realise that the future of the human-kind is deeply fractrured. Various tributaries like movement for disarmament and world peace, anti nuclear movement, campaign against reductionist and patriarchical science, feminism, green philosophy, population explosion, the Gaia movement have made the environment movement a surging stream. And today the environment has developed deep philosophical roots bordering almost on mysticism. Frequent references are being made in the deep ecology literature to Buddhism, Taoism, Gandhian philosophy and the concept of Brahma.[7, 8, 9]

Zero rate of growth, even a degree of deindustrialisation is being advocated by prominent ecologists like Goldsmith who has been in the forefront of the environmental movement for many years. While environmentalists are deeply concerned about the quality of the life in overdeveloped countries, for the people of less developed countries the problem is that of life itself. Hence call for straight-away deindustrialisation or zero rate of growth for the poor countries seem inappropriate and a development model which could be sustained for both rich and poor countries should be the ideal.[10]

On the other side of the coin the economists have also travelled a long road in the meanwhile. Upto fifties and sixties the over-riding concern was growth and growth and growth. The prevailing mood was more the merrier, bigger the better. Then came balanced growth, inter sectoral and inter-regional, thereafter economists got concerned about growth with equity and social justice. And when inflation raised its ugly head, the call was growth without inflation and today the call is for sustainable development i.e. development taking environment into consideration or development without destruction.[11, 12]

The figure No. 1 explains divergence and convergence of views on environment and development.

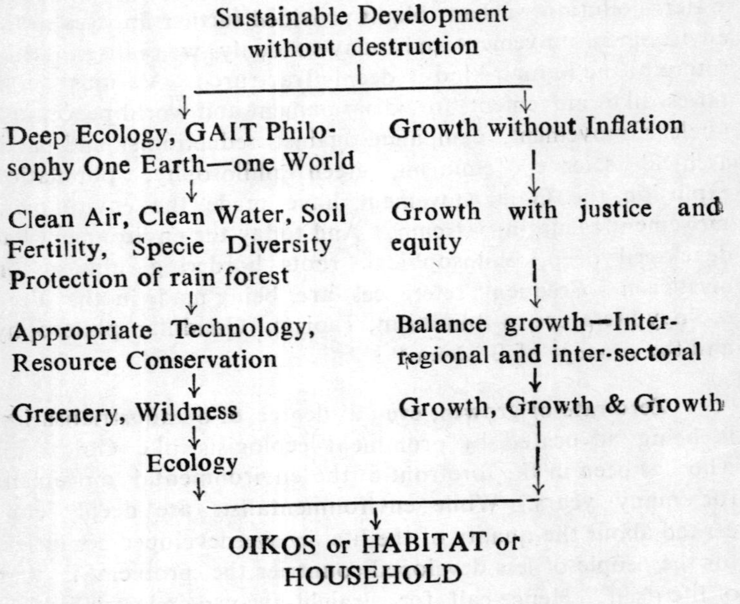

Figure 1

It has been amply recognised that underdevelopment of the third world is directly linked with the over-development of the first world. The extravagant life style of the people of the North has become possible because of the global exploitation of resources and capital intensive technology. Many of the third world environment crises can be traced to the profligacy of the rich nations. Notwithstanding brave statements like, if it is very easy to substitute other factors for natural resources then there is, in principle, no problem. The world can, in effect, get alone without natural resources, so exhaustion is just an event, not a catastrophe,[13] by Mr. Robert Solow, the noble prize winner in economics, 1987, the fact remains that the economic engine of the over-developed countries operates with plundering of resources on a global scale. Gandhi's statement in this context seems very relevant. When asked if India would pursue the western model of development after independence, Gandhiji said, "God forbid that India should ever take to industrialism of the West after the manner, the economic imperialism of a single tiny island kingdom (England) is today keeping the world in chains. If an entire nation of the 300 million took of similar economic exploitation, it would strip the world bare like locust."[14]

In the past on countless occasions, the west has flexed its military muscules to have its share of the third world natural resources and even today it is flexing its economic muscules and sometimes indulging in nuclear sabere rattling when a steady supply of natural resources is threatened.[15]

The technology used to sustain the high mass consumption Society of Rostow is highly energy intensive and Capital intensive to Ethics of capitalism and unabashed pursuit of acquisition enlarged the scale of production and also led to centralisation of production and distribution process. This process created a kind of technology which radically altered our outlook towards nature. 'Mother' Nature, respected and reverred, turned to be 'female' nature, to be conquerred, subjugated, dominated, raped, exploited and mined.[16] The attitude that nature could be pillaged and plundered, tamed and overpowerd by humans with impunity gripped the homo economicus. Since the socialist countries adopted the same western technology, their attitude towards nature did not differ. No wonder both capitalist and socialist countries are facing nature's backlash.

"Nature which once terrified men by her 'incomprehensibility', now frightens man by its infirmity and fragility of its intimate mechanisms. The fear experienced by man in the past before alien, terrible nature has given way today to fear of the uncoordinated powers of modern industry. The whole process of interaction of man and nature and of society and nature has begun to acquire paradoxical features... The tamer and the tamed are quite understandably in a dichotomy. And since nature is not passive at all, she takes revenge on the man who scorns her laws.[17]

To understand natures' magnitude of revenge it is necessary to know the course nature itself took to create this wonderful world. It took millenia, in fact, billenia for the life-support system of the earth to evolve and myriads of living being to blossom.

Much of our water, (97.3%) craddle of life (97.3%) is in oceans and it is salty (3.4%). Any increase in saltiness will eliminate most of our oceanic plankton and other forms of life from our oceans. That will seriously alter oxygen-carbon-dioxide composition of the air. Our fresh water is acidic and increase in acidity will also severely affect the water quality affecting crops and living being.

Air we have has about 21% oxygen. Oxygen is vital for our life, but a step-up in oxygen can make itself combustible. CO_2 is present in air in very very small quantity (0.03%). This provides some green-house effect which is just necessary to keep the world warm enough for the myriad life farms to exist and flourish. Just double the amount and even then it is an infinitesimally small quantity and then the 'green-house effect' increases to a point leading to unforeseen climatic changes and global catastrophe. It is about this prospect we are so much worried. The world sustains such an excellent diversity of living organisms in only a narrow band of temperature.[18]

Nature has taken billions of years to create the protective ozone layer in the stratosphere. This layer of ozone filters much of the deadly ultraviolet-B rays being constantly bombarded by the sun. Without this layer no living organism could continue to exist. 'We will be fried to death', says an environmentalist.

Similarly millions and millions of years passed before the thin top layer of soil was formed. It is this thin layer of 10 to 12″ of top soil which provides food to 5 billion plus people today and will continue to sustain a larger number of people and other living organisms. If this layer goes away, we may have land, but not soil. Human civilisation will wither away.

And take the vast web of life each integrated with the other in various ways not actually fully understood by us. Remove one specie and how the living world will be affected is difficult to comprehend. How precarious is our existence then ? How delicate is the balance ? If this balance has continued till today, it is because nature observes some laws rather strictly.

What is development doing to this balance ? Millions of factory chimneys and automobiles are spewing billions of tons of carbon dioxide, carbon monoxide, oxides in nitrogen and other noxious gases into the atmosphere. A number of other chemicals are also being injected into the atmosphere in enormous quantity. Because of atmospheric circulations there pollutants are transported to long distances and their concentrations are depleting ozone layer even over Antarctica.[19]

Because of the property of the CO_2 to trap solar heat, greenhouse effect is steadily increasing and the globe is warming up. Low lying countries like Holland and Maldives have started apprehending their submergence because of rise in sea level which may take place as a result of melting of ice-caps as global temperature rises.

London, the pace settler of industrial revolution is likely to be submerged in the next century.[20] Major metropoliges like Toky, New York will also be threatened. Lester R. Brown, President, World Watch Centre predicts a global food emergency because of unforeseen climatic changes.[21]

When part of these gases turn to acids, they cause acid rain. Today thousands of water bodies in scandinavian countries have become virtually dead because acid rain has made water very acidic. Soil also is turning acidic reducing its productivity.

Our great rivers have sustained civilisation for centuries. River basins have acted as carddles of many a civilisation. They have acted as our arteries. Today because of mindless dumping of wastes, both human and industrial these rivers once the nourishers of life have become sewers. This is true from the Ganga to Rhine. Great oceans have become great sinks and many oceanic living organisms including sea weeds and planktons are threatened under the impact of oil spills and other contaminants of all kinds—chemicals and nuclear.[22]

Forests which provided a natural cleansing mechanism and maintained the oxygen and Co_2 balance apart from providing habitat to countless varieties of life-forms are disappearing at a fantastic rate. Tropical forests are disappearing at a rate of between 7.6 million and 10 mn. hectares and thousands of species are becoming extinct along with disappearance of forests.[23] This is altering rainfall pattern and ground water recharge. As regards land some 29% of the earth's land area suffers light, moderate, or severe disertification; an additional 6% is classified as extremely severely desertified.[24] Top soil, precious for food security is washed away because of bad management. Besides all this, think of nuclear radiation, and chemical invasion caused by thousands of chemical products already in market or being introduced every year.

The life-support system of the world evolved over billenia is under real threat. The plant is in peril. Today awareness about this is spreading fast, but what is necessary is awareness about the root causes-exponental growth in population, energy consumption, inappropriate technology and above all appropriation of world's, resources by few to maintain an unsustainable and vulgar life-style. People will have to learn that they should live and allow the future generation to live too.

The awareness that the earth does not belong to humans, but the humans belong to the earth, in the words of the Chief Seattle, USA has to grow. Both the anthropocantric and the ego-centric world view should give way to an eco-centric or bio-centric would view.

But what should the poor countries do ? They are late comers in their pilgrimmage to the temple of prosperity. Development in the sense that dire-destitution, hunger, ill-health, shelterlessness, inadequate clothing etc. should go is yet to take place. In the plan parlance basic needs of millions of people are yet to be met.

Let us take food first. Food adequate quantitatively and qualitatively could be provided through so-called 'scientific farming'. This scientific farming depends on energy, synthetic, fertiliser, insecticides, pesticides, herbicides, HYV seeds or mechanisation, chemicalisation and hybrydisation. To use the words of Fritz of Capra, modern agriculture has shifted from soil to oil. Synthetic fertiliser, particularly nitrogeneous fertiliser is produced mostly from fossil fuel whose supply is finite. Use of fertiliser pollutes air, water and soil. Excess fertiliser application affects food quality too. It makes crops more prone to diseases and pests requiring application of pesticides etc. Pesticides world-wide have proven worse than pests. They too poison water and soil, eliminate pest predators. More than that food is poisoned by pesticides affecting our health. Many of these chemicals are carcinogens and mutagens. Irreparable damage is being done to our genetic heritage by agro-chemicals.

The hybrid seeds ensure higher production. But with rise in input costs the margin is getting narrowed. Poor farmers cannot afford without huge state subsidy. Production of fertiliser requires non-renewable resources, capital and energy and its application has adverse environmental consequence.

Use of hybrid seeds is leading to decline in genetic

diversity which used to provide insurance against diseases, insects and pests, droughts and floods. This genetic erosion may cause disaster in future. The Irish potato famine in the last century which led to the death of about 2 mm. people was caused because of a narrow genetic base of potato crop.

The other important input is water provided mainly from huge dams built at enormous public expenditure. Canal water is leading to waterlogging and salinisation of soil. In many areas productivity has started declining in fact.

The present agricultural strategy might have brought 'Green Revolution', might have trebled food production, but in the long run ecologically speaking consequences will be disastrous.

On the other hand, bio-farming with emphasis on in situ water conservation, bio-fertiliser, bio-control of pests, species diversity, crop rotation is less commercial energy intensive, cheap and clean. It is ecologically more benign than the modern agriculture which is destructive. About productivity? a recent study undertaken by the prestigious US Academy of Sciences reported by the Times of India says :

1. The findings of the council could have important consequences not only for U.S. agriculture, but for farming around the world. It appears to go against the conventional wisdom and widely prevalent agricultural policies which have discouraged farmers from trying natural techniques.

2. For decades, farmers around the world have been taught that the best way to increase crop yield is to use plenty of chemical fertilisers and pesticides. In many countries, including India, supplies of such chemicals have been subsidised by the Government, though in recent years, environmentally conscious groups have strongly argued that natural farming techniques are equally good for farm productivity and positively better for the environment. Their

cause is likely to receive a shot in the arm with the publication of this report by a highly respected U.S. science body.

3. Natural farming is not a single method but rather a variety of techniques, according to the report. These techniques have the common goal of reducing costs, preserving the environment and protecting human health by lowering or eliminating the use of toxic farm chemicals and drugs. Natural farming usually requires harder work and greater management skills than chemical based agricultural practices.

4. The study, which began in 1984, cites examples of farms around the U.S. which have developed effective natural methods of cultivation. In a dairy, cattle and grain farm in Knox country, Ohio, for example, natural practices and a conscious eschewing of fertilisers and pesticides for more than 15 years, has produced corn yields that are 32% higher than the average for the country yields of soyabeans to have been 40 per cent higher.[25]

Will we be less developed if we grow our food in a natural way than the chemical way ? Natural farming has been experimented in Japan and Masanobu Fukuoka, the Magasaysay award winner has proven that food could be grown in natural way with very high productivity which chemical agriculture may fail to attain. His logic is : All of us want good health, good health we can get by taking healthy food, healthy food we can get from healthy soil, and healthy soil we cannot get by poisoning our soil by chemicals.[26]

With natural farming the poor countries can get rid of the strangehold of the international agri-business interests. Huge State subsidies for fertiliser and pesticides could be dispensed with leaving money for other sectors. Productivity could be sustained over a long period with least external inputs.

Clothing is the next important human need. People could be clothed better by cloth produced by huge automatic energy-intensive textile mills located in cities. Cotton has to be produced and transported to cities and manufactured cloth again has to be transported back to lakhs of villages. We need commercial energy to produce and transport. We also need people to leave their villages and trek to cities. On the other hand cloth could be produced in a decentralised manner, in thousands of villages. Are we less developed if we are clothed by handloom clothes ? Will mill-made cloth be really cheap if all social costs are taken into consideration ? If pollution control measures are taken at all stages of production ? Production process of synthetic clothes is polluting and synthetic dyes also are polluting. Are we less developed with cotton and white clothes or clothes coloured with natural dye ?

The other important basic need is housing. Asbestos roofing is dangerous and environmentally not permissible as it causes cancer. RC roofing depends on resources that are non-renewable and energy intensive. It is capital intensive too. It will be an impossible dream for our planners to provide concrete structures to millions of homeless or inadequately housed people. Conventional housing could be improved with improved mud architecture. Traditional housing designs with emphasis on local resources like mud, laterite blocks and thatch are not only cheap, accessible, also they keep homes cool during summer and warm during winter reducing energy requirement.

In the health sector the present emphasis of the planner is on curative aspects, sophisticated diagonistic equipments and professionalism. It has been estimated by the WHO and UNICEF that 75 to 80 per cent of ailments, mostly water borne could be eliminated by improved sanitation and safe drinking water facilities. In India about 15 lakh under fives die every year or on average little less than 3 children die per minute before they celebrate their fifth birthday. Diarrhoea, dysentry, typhoid and *hepatitis—all* water-borne are taking away precious lives.

Enter into a village in rainy season, and it is a nightmarish, nay, a night soilish experience. Lack of sanitation and safe drinking water is probably the greatest environmental problem, with a number of ramifications. Cause of environment and development is served best if sanitation situation improves.

The other basic need for millions of poor people in the third world is firewood. Since their very existence is bio-mass oriented[27] food, fibre, fodder housing, fertiliser, medicine etc. it is necessary that our forests are protected, regenerated and rehabilitated and plantation undertaken.

It is true that wood is no substitute for food, but fire food is the principal fuel to prepare food for the vast majority of the people of the third world. Shortage of fire-wood has become the most important form of energy crisis. 'Barren slopes and cold hearths' go together.

Besides this consequences of disappearing forests siltation of dams, recurring visitations of flood and drought, decline in underground water, sources etc. have been well recognised by now. It has also been acknowledged that further expansion of agriculture at the cost of forests will only be counter-productive.

Intense heat and humidity have produced tremendous diversity of plants and animals and our forests are storehouses of genetic diversity. If genetic manipulation has produced a variety of new seeds with higher productivity, resistance to drought and pests etc. we will have to go to tropical forests in search of wild relatives of the cultivars for further improvement[28].

Commenting on burning of the Amazonian forests the Economist writes in its editorial.

"As they blaze some of natures' miracles vanish for ever.

Brazil contains nearly one third of all the tropical forests and a greater variety of plants than any other country. The Amazon has more types of fish than in all other rivers of Europe, the trees are home to more species of birds than in all the forests of North America. Somewhere in the Brazillian jungle may grow cures for cancer or AIDS or malaria. By allowing the destruction of this cornucopia, Brazil has acquired a reputation-especially among the young-as the worlds' environmental thug."[29]

It writes further, "The roads into Amazan, which have opened it to damaging settlements have generally been built with the help of first world donors; the timber taken out has found its way into the first world sitting rooms."...[30]

Mammon as much as nature should persuade Brazil to conserve the Amazen. Much deforestation results from activities fostered by bad fiscal policies; from cattle ranching designed to mop up tax losses; from unprofitable hydro electric schemes built to provide underpriced electricity, from iron smelting built with lavish state aid, from small farmers displaced by subsidised mechanisation of big farming in the South."[31]

"Brazil could make a better living from keeping its forest and exploiting it sensibly than from destroying it. Botanists turned economists studying the value of forest products in local markets, recokon that fruits, nuts, rubber may be worth more than thrice as much as the same land logged or turned into cattle pasture."[32]

What is true in regard to Brazil is true with regard to many other poor countries having natural forests. Should then we continue to raise the controversy of man vevrs. monkey ?

The other important concerns of developmental planning in backward countries are social equality and justice and full employment. Decentralised production, environmentally appropriate technology, a simpler life style, wasteland development and massive afforestation programme ecologically sustainable agriculture can generate employment on a large scale, halt rural trekking to urban slums and promote social equality. Provision of safe drinking water and improvement of sanitation can mean longer lives for the children and that can make the task of pursuading families to have smaller families easier apart from making our habitat more liveable.

To conclude for poor countries ecological imperatives serve the developmental objectives best and people of over-developed countries may live simply so that other people simply live.

References

1. Jayanta Bandyopadhyay and Vandana Shiva, Political Economy of Ecology Movements, Economic and Political Weekly, Vol. XXIII-24, June 11, 1988.
2. Charlme Spretnak, 'Eco-feminism', Eco-Spirit, 7, No. 2.
3. Arne Naess, Deep Ecology and Ultimate Problems, The Ecologist, Vol. XVIII, No. 4/5, 1988.
4. Zafar Futehally, Conservation & Development, India International Centre, Quarterly-Environment India, Vol. IX, No. 3-4, 1982.
5. Robyn Eckersley, The road to Ecotopia? Socialism Vrs. Environmentalism.
6. Rajni Kothari, Environment and Development in Asia and the Pacific, Experiences and Prospects, UNEP, 1982.
7. Briantokor, Social Ecology, Deep Ecology and the future of green Political Thought.
8. Richard Sylvan and David Bennett, Taoism and Deep Ecology.
9. Edward Goldsmith, The Wey : An Ecological World View, The Ecolcgist, Vol. XVIII, No. 4/5, 98.
10. Edward Goldsmith, the Editor of the Ecologist, The Great U-Turn ... Deindustrialising Society.
11. Our Common Future, The World Commission on Environment and Development or the Brundtland Report.
12. Robert Riddele, Eco-Development, Gower, 1981.
13. Robert Solow, The American Economic Review, May, 1974.
14. M.K. Gandhi (1928), Young India, December, 20, p. 422, Quoted by Jayanta Bandopadhyay and Vandana Shiva.
    —Political Economy of Ecology Movements.
    —Economic and Political Weekly, June 11, 1988.

15. Robert Ridell, Eco-Development, Gower, 1981.
16 Vandana Shiva, Staying Alive, Women Ecology and Survival in India, Kali For Women, New Delhi-1988.
17. Ilya Novik, Society and Nature, Progress Publishers, Moscow.
18. Patrik Rivers, The Stolen Future : How to Rescue the Earth for our Children, Merlin Press, London.
19. Tomoyoki Ito, Antarctic Submicron Aerosols and Long-Range Transport of Pollutants, AMBIO, Vol. XVIII, No. 1, 1989.
20. G.P. Hekstra, Global Warming, Rising Sea Levels, The Ecologist, Vol. 19, No. 1, 1989.
21. Lester R. Brown, Climate Change and the Food Outlook, Populi, Vol. 15, No. 4, 1988.
22. Our Common Future, World Commission on Environment and Development, p. 264, Oxford University Press, 1987.
23. *Ibid*—Chapter-6.
24. *Ibid*—Chapter-5.
25. ***Times of India***—Sept. 19, 1989, Bombay Natural Farming Better, US Report.
26. Masanobu Fukuoka, The Natural Way of Farming, Japan Press.
27. Anil Agarwal, Citizens' Report, Centre of Science and Environment, New Delhi.
28. Genes from the Wild-Earth Scan, London.
29. The Economist, 9 September, 1989.
30. *Ibid.*
31. *Ibid.*
32. *Ibid.*

12

Environmental Degradation in the Mayurakshi Catchmant, Bihar

*Miss Nandini Chatterjee**

Of fundamental concern in recent planning policies oriented towards the sustained development of resources, is the emerging environmental crisis. Unless one takes cognizance of the multifarious ecological imbalances, the concept of 'Sustained Yield' of water resources in major river valley projects, also, will continue to remain a nurtured dream. *Prima facie* progressive deforestation succeeded by unscientific land use practices associated with accelerated soil erosion in catchment areas, culminate in increased sedimentation rats of reservoirs. Thus the identification of these problems needs to be given high priority in watershed management policies. Giving due consideration to these facts, the main focus of enquiry in the present paper has been the causes, nature and consequences of environmental degradation in the catchment area of the Mayurakshi River Valley Project. As such this is one of the three major river valley projects of the State of West Bengal, lying east of the Santhal Parganas district of the State of Bihar. There is thus the political issue of the catchment area

*Senior Fellow, CSIR, Calcutta University, Calcutta.

lying in one state and the major part of the canal command area lying in another.

The Study Area

The Mayurakshi River Basin extends over 8,805 sq. km. in the districts of Bihar and West Bengal, India. Since 1954 this basin has been brought under the Mayurakshi River Valley Project with the prime objective of irrigation and associated hydel-power generation. Flood control is incidental. In general, this river basin is located over a drought prone, plateau/plateau fringe surface, characterized by hilly terrain along the upper reaches, grading down to gently undulating plains in the east and south east. In fact this is the continuation of the extensive Chhota Nagpur Plateau of Bihar.

The broad geomorphic units that have been identified are :

i. Marginal Plateau Surface with Hills (> 200 m)

ii. Plateau Fringe Zone (200-100/75 m)

iii. Old Alluvial Tract (upto 30 m)

iv. Recent Alluvial Plains (< 30 m).

The catchment area consists of the first zone and adjoining parts of the plateau fringe zone. The latter may more precisely be termed as the piedmont zone. Strictly speaking the catchment area above the Mayurakshi Reservoir extends over 1859 sq. km of which the reservoir occupies 67 sq. km. In the present study, the upper reaches of the other major rivers, viz., the Brahmani, Dwarka, Siddheswari and Bakreswar, have also been studied to provide a more comprehensive picture. It may be mentioned that barrages have been constructed across most of these rivers. Hence a study of their upper reaches is also necessary.

Geologically, the catchment area is predominantly formed of Archaean gneisses, mainly granite gneiss and hornblende-biotite gneiss. Coarse grained granite forms the domal hills.

Trikut Hill (753 m), which forms the source region of the Mayurakshi river, is one such example. Structurally coarse grained granite forms the anticlines whereas biotite gneiss forms synclines (GSI, 1964-69). The region is a highly metamorphosed and deformed peneplained surface. Having been subject to continuous uplift, the region is much faulted and fractured. This is evident from the linear stream patterns and occurrence of hot and cold springs. Pegmatite and quartz veins, often brecciated, form conspicuous rocky knobs. Dolerite dykes also occur. Strewn boulders, sheet rocks and stony wastes are also conspicuous features.

The Lower Gondwanas are represented by the Talchir Series comprising of green laminated shales and soft fine sandstones and the overlying Barakars with soft massive sandstones and shales interbedded with thin coal seams. They form patches of low lying plains in the west and north east. The Upper Gondwanas are represented by the resistant, coarse-grained Dubrajpur sandstones forming low hills with steep slopes and cliffs. This zone is also highly faulted. The Dubrajpur sandstones are oftem ferruginous in nature. The Gondwana sedimentation culminated in the effusion of Rajmahal Volcanics consisting of 10 horizontally bedded basalts and 4 inter-trappean beds of tuff, clay and ash. They constitute the Rajmahal Range of Bihar. The hills in the northern part of the catchment area form the southern extension of this Rajmahal Range. Heights rise upto 502 m capped by laterite formations, insitu. The region is very highly faulted with a concentration of springs along the boundary of the different series. Even in the adjacent piedmont plains, in the southeast, where trap rocks prevail, these springs are observed. Another conspicuous feature is the pevalence of stony wastes.

In general, soils vary from red sandy soils to sandy loams with laterite cappings over basalt. Undefined channels are characteristic drainage features. The streams are mostly seasonal in nature flowing only during the monsoons. The rest of the year they form dry water courses. Broken banks and gullies occur but the latter are not very extensive.

Structural control of the major streams are apparent in their angular bends and straight reaches.

The average annual rainfall amounts to 1420.56 mm at Dumka. About 80% of this rainfall is concentrated during the monsoon months (June/July to September). October to February constitute the winter season whereas March to May form the summer season.

Considering overall landuse, about 65% of the total area is under cultivation, 6-7% under forests and about 23% under cultivable wastes. The region has long experienced human interference.

Parameters of Environmental Degradation

The basic problems that have been disturbing the environmental set up are discussed below :

Deforestation

As early as 1803 Gastrell reports, "There can be no doubt that the unrestricted clearing of the Jungles has had great influence in decreasing the annnal amount of rain and impoverishing the country. This bears a testimony to the emerging ecological imbalance that has with successive years increased in magnitude. Lack of data on the past extent of forests has necessitated reliance on theoretical information. By the beginning of the 20th century, most parts had been cleared for cultivation, confining dense forests along the hills, steep ravines and slopes. In certain areas, even these regions were not spared from the reckless exploitation. The basic cause was the unscientific practice of 'jhum' or shifting cultivation by the Santhal tribes as well as the prevalent practice of selling firewoods to the towns. This problem was further intensified by cattle grazing which restricted the coppice from growing again. The predominant species was the sal (*Shorea Robusta*) but along the higher slopes mixed forests existed along with bamboo. From sal and pial (*Buchanania latifolia*) along the lower slopes, variety increased to murga (*Pterocarpus*

Marsupium) satsal (*Dalbergia Latifolia*), etc. On the jhum lands the original forests gave way to shrubby undergrowth. The only exception was the Old Reserve in the Dumka region extending over 93 sq. km situated between elevations of 185-520 m. with sal as the chief specie (O'malley 1910). Mention may be made of the mohua (*Bassia Latifolia*), palas (*Butea frondosa*) and asan (*Terminalia tomentosa*) used for successful lac culture. Even till 1870's these forests were abound in wild life such as tigers, leopards, deer, wild elephants, etc. The last wild elephant was shot in 1893. Deforestation along with the hunting habits of the Santhal tribes were responsible for the extinction of the fauna.

At present, forests cover a negligible area. Thick cover is recorded in the Mahuagarh Hill (502 m), in the catchment area of the Brahmani River, where forests cover about 25% of the total area. This is followed immediately by the hills around the Masanjore Reservoir where values reach 15%. Elsewhere values are mostly 2-5%. Comparison of old and recent topographic maps depict a constantly decreasing forest area, even in the Mahuagarh hills, by about 50%. The source region of the Mayurakshi, i.e. Trikut Hill (753 m) has a meagre forest cover at present. Other isolated gneiss hills were also recorded. The predominent species is sal with a decreased area under mixed jungle, occurring in small dispersed patches. Afforestation with plantations of eucalyptus and akashmoni (*acacia auriculiformis*) along with a negligible amount of teak have been introduced. The original girth and height of the existing sal forests have been considerably reduced. Offer, they are replaced by mahua and palas in the late stage of retrogression. Wild game has almost disappeared. Open jungle and scrubland are conspicuous features.

The long run ecological effect of afforestation with eucalyptus and akashmoni is still a matter of contention. No doubt, they are the most rapidly growing species hence their planting, to combat deforestation, may be justified. Yet, it has been noted that their consumptive use of water and soil nutrients is alarmingly high. This may have serious repercussions on the groundwater table and soil fertility in the long run. It has been estimated by UNESCO that eucalyptus hybrid needs 1200 mm of rainfall. In this drought prone area the trees satisfy the demand from the water table. Moreover they intercept less rainfall due to their structure. Human interference, through removal of leaves shed by these trees, reduces organic matter of this soil. Lack of social awareness may lead to a failure of afforestation programmes. During field surveys, it has been observed that villagers often stealthily make cuts in the tree trunks and apply certain chemicals which cause the trees to dry up rapidly, making it convenient to cut and transport the timber. Necessity of fuel has been one of the major factors causing deforestation. At present, small plots of government owned wastelands are being allocated to landless families for tree planting. They can not own the land but are given usufruct of the forest.

Impact on Rainfall

A decreasing trend of rainfall is conspicuous in the long run During the 1800s the average annual rainfall was 1534 mm which had decreased to 1389 mm in 1960. Even in the police station of Siuri bordering the catchment area, for which yearly rainfall values were available for the period 1930-1985, linear regressions analysis revealed and yearly decrease ($Y = 1219.64 - 0.654X$).

Increase of Erosion

This has been a direct consequence of deforestation and unscientific land use. Study of topographic maps and landsat imageries along with field surveys reveal certain interesting features. Cliffs and rocky slopes for instance are founds to be more extensive at present as along the Trikut Hills. Bare rocks are conducive to increased weathering and erosion. The summit of the hills have become more dissected. Isolated rocky knobs have developed in certain areas.

In general the piedmont region is characterized by frequent occurrences of rocky knobs in succession, implying rock or dyke outcrops, as well as strewn boulders and stony grounds. However, compared to 1920s, a sequential development is in progress. By 1980s those areas with numerous rocky knobs have been transformed to strewn boulders, sheet rock and stony ground. Rocky knobs are found to be developed over quartz and pegmatite veins besides gneiss. Some of them attain heights of 15 m.

During the 19th century there existed one small cascade along the bed of the Brahmani River at Singhpur, where it traversed an extensive bed of basalt. The fall was 3 m (0' Malley, 1910). Today this has been eroded down to basalt boulder bed.

Impact af Quarrying, Mining and other Land Uses

Recent topo-sheets and aerial photographs reveal a concentration of extensive stony wastes along the basaltic terrain. Sudden appearance of numerous spring lines along a juncture testify to the occurrence of faulted trap rocks either exposed as hills or as plains with a lateritic or alluvial veneer. Field checks have been carried out along this region. It has been observed that most of these stony wastes have their genesis in the extensive quarrying of basalt trap rocks. Alongside these quarries, stone-chips are prepared. Old maps revealed these sites as cultivated lands. The fact can not be refuted that human interference has induced this progressive exploitation of cultivable land. The presence of these basalt quarries were recorded in the past. They were used for building temples, forts, and other structures in the vicinity of the hills and in a more restricted manner for road metal. Lack of transport facilities rendered their export economically unremunerative. It was only within the proximity of the newly developing railways there was mushrooming growth of such quarries. This trap rock also yielded agates and chalcedony and other rock crystals. High demand for building materials along with development of transport facilities have augmented quarrying functions.

These quarries however, provide employment to the villagers as they function throughout the year. In contrast, the drought prone rocky terrain inhibits cultivation during the winter and summer seasons when the lands lie fallow. Previously these villages used to cultivate the rainfed aman rice, the main crop of Bihar. Domestic water supply was provided by dug wells tapping the upper weathered mantle and underlying fracture zones, if the depth of bedrock was shallow. Some of these dug wells are still in use. However the opening of these quarries have tended to disrupt the groundwater balance. Groundwater is often encountered at shallow depths and is pumped out either for use in the subsequent processes or allowed to run down. Extensive gullies have been initiated along the banks of these pits, which unless checked, will tend to encroach upon the neighbouring agricultural lands. Another serious fact to be considered is that once the basaltic pillars have been totally removed, there will not be sufficient soil to fill up these pits to restore the landscape. The adjacent areas used for stone chips processing creates another problem. The original fertile cultivable land is veneered by dust and minute fragments, These dust particles often move to the neighbouring zones. As these quarries are in temporary lease, no measures are undertaken to prevent dust pollution of the villages. The ill effects may become obvious in the long run. But at present the villagers are totally oblivious of these facts and are somewhat satisfied with the jobs.

Coal mines and local iron-smelting units were a characteristic feature of these areas even as early as the 19th century. Coal mining pits occurred in abundance along the beds of the Brahmani and Siddheswari rivers which traversed Barakar measures and Talchir series (Gondwanas) which are the chief coal bearing beds. But the coal in this area is of inferior quality. Continuous exploitation occurred with development of Eastern Railways, which linked Calcutta with the major coal belts further west. Today most of these mines have been exhausted because the coal beds were not extensive. Similar was the case of the iron smelters which worked on the local iron ores, occurring in the seams, disseminated among and spreading in a tangled manner through the trappean claystone. It is an infiltrated oxide of iron deposited in the fissures and joints of the trap rock, with beds only 1.5 m thick. Thus the ores of one place were soon exhausted and the site shifted, leaving villages with neglected slag heaps. The main villages were Belia Narayanpur, Deocha, Dhamra and Ganpur with 30, 30, 4 and 4 furnaces respectively. Each was capable of producing about 3/4-1 ton of pig iron in a week, combined production being 2380 tons per year (GSI, 1851-52).

Hence the deterioration of the land was initiated as early as the 19th century with the formation of cultivable wastes by the slag heaps. Today there sites have been discarded, thus giving a misleading picture of the man-made landscapes. Unscientific cultivation practices also have their repercussions. The deforested lands are not cultivated throughout the year due to lack of irrigation facilities. These fallow lands are more susceptible to erosion. Often the superficial soil veneer is washed away exposing the underlying stony horizons. This reduces soil fertility. However, the villagers have tried to combat this through terraced farming. In recent years the Government has taken initiative in introducing soil conservation measures.

Impact of Erosion on Drainage

All these factors have induced soil erosion and consequent changes in the drainage patterns. On a localized scale gully erosion has been initiated along the edges of quarries and coal mines. In the foothill regions, there has been an increase in the number of undefined channels. The superficial colluvial and alluvial veneer, more or less coarse textured, have caused the hill streams to either disappear at the foothills or form undefined channels in this reach. The latter may occur due to the presence of an impermeable substratum at shallow depth or a locally high groundwater table thus sustaining some flow during the rains. The disruption in the soil structure in and around the collieries have resulted in a conspicuous reduction in drainage density and stream frequency. Over the Talchir series, for example, the fine textured shales had initially produced a fine trellis drainage pattern. With removal of shales and exposure of underlying sandstones or by the bringing down of coarse alluvium from neighbouring regions, a coarse texturedic pattern has emerged. This may also develop with the removal of the finer surface soil and exposure of coarse, stony soils. There has been an extension of broken banks along some of the rivers.

Thus the surface water resource is visibly influenced. This may in the near future affect the water regime of the major rivers.

Increased Sedimentation in the Reservoir

Disruptions in the ecological balance, through deforestation and soil erosion, has culminated in increased sedimenta tion rate of the Mayurakshi Reservoir at Masanjore. This is evident from past and present data recorded by the River Research Institute, West Bengal as well as field surveys. The siltation index which was 13.56 ham/year/100 sq. km. in 1962 rose to 20.62 ham/yr/100 sq. km. in 1972. Since then afforestation and soil conservation measures have slightly reduced this to 16.42 ham/yr/100 sq. km in 1980. With respect to the ratio 'n' of Actual Silt Index (A) to Designed Silt Index (D), the Mayurakshi reservoir has a value of 4.56, emerging the third highest amongst the 12 major river valley projects of India. The original capacity of the reservoir was 60,797 ham in 1954. Two reservoirs, one the original reservoir behind the Canada Dam and the other a freshwater lake occurring in the northeast beyond the highway have been observed. This natural lake has been formed by the confluence of streams in the neighbouring hills. Water from this lake is allowed to flow into the actual rerervoir. The contrast between these two water bodies bears testimony to the turbidity of the reservoir. Landsat imagery for 1987 also supported this fact. Only negligible amount of marginal areas of the Mayurakshi reservoir appeared to be less stagnant. About 6.5-7.5 km. upstream

	Ca ++ as $CaCo_3$	Mg ++ as $Mg(OH)_2$	SO4—as $(NH_4)_2SO_4$	Clas Nacl	%OC	Fe_2O_3	pH
	←——	—— ppm ——	——	——→		in ppm	
Reservoir	110	8.0	4.0	11.0	—	5.0	7.3
Lake	13	trace	trace	trace	—	—	7.2

the density was 1.003, pH value 7.2, viscosity in centipoise 0.806, % salinity .0124-.0152, % suspended solids .0096-0180. The maximum turbidity zone lies slightly upstream of the dam. Chemical analysis reveals the following character :

Conclusion

The environmental implications emanating from this study can not be ignored. The increasing ecological imbalances are clearly discernible, with the passage of time. Another feature is their intricately linked nature. In fact environmental degradation is the outcome of multiple processes rather than an isolated phenomena. The visible benefits accruing from the implementation of river valley projects often leads to an overall complacent attitude. It is only when the ecological crisis assumes recognizable proportions, that hue and cry is raised. Afforestation programmes, soil conservation measures, scientific mining practices and a creation of social awareness amongst the villagers are just some of the palliatives to face this problem. However, the vital issue is watershed management. Often this problem is left untouched as one department shifts the responsibility to another and vice versa. A similar situation arises in case two or more political units are involved. If the command area lies in one unit and the catchment area in another, the non-benefiting state develops a laissez faire attitude towards tackling these problems. The benefiting state, also, is reluctant to initiate measures without cooperation. This can be solved only through consolidated efforts and mutual understanding of the various departments and states involved.

Acknowledgement

The research worker is grateful to her supervision Professor Bireswa Banerjee and also thankful to the CSIR, India for financing research.

References

Gastrell, J.E., (1803) –West Bengal Forests, Centenary Commemoration Volume, Forest Directorate, Govt. of West Bengal, pp.

Hunter, W., (1876)—A statistical Account of Bengal, Vol. Rep., by D.K. Publishing House, Delhi, 1974, pp.

O'Malley (1910)—Bengal District Gazetteers : Santhal Parganas, Rep., by Logos Press, N. Delhi, 1984, pp.

G.S.I., (1964-69)—"Systematic Mapping of parts of Santhal Parganas district, Bihar"; Unpub., Field Report for season 1964-69, Geological Susvey of India (GSI)., by U. Prasad.

13

Ecosystem Approach for Sustainable Development

*Kumar B. Das**

All economic activities necessarily exert a claim on environment both for the input sector (resource base of economy) and for the output sector (the effluents and residuals from economic activities). The first claim leads to exhaustion of natural resources, the second one to pollution and environmental degradation. This chapter seeks to highlight the problem of environmental deterioration and depletion of energy potential. It analyses the unintended consequences of intended economic action in an analytical framework. While analysing the future prospects of the supply of energy and environmental services the author tries to make out a case for having an ecosystem approach of an intergenerational set of criteria to the resource management.

The Interdependent System

Ecology, Energy, and Economy are highly intertwined

*Professor, Department of Applied Economics, Utkal University, Bhubaneswar.

into complex systems (Fig. 1). The ecology is the biology of ecosystems, which refels to the interaction and interdependence, an orderly working totality of complex factors constituting the total environment. The natural ecosystem consists of the physical environment inhabited by plants, animals, each of which in turn modifies the climatic, the hydrological and nutritional aspect of the environment. The interacting men, animals, plants and the energy flow of the environment make up the ecosystem[1]. The ecosystem functions as a complex of interacting systems involving the living elements and non-living. The figure 1 illustrates the interaction and interdependence between economy environment, energy and political decision making of the country or the world as a whole.

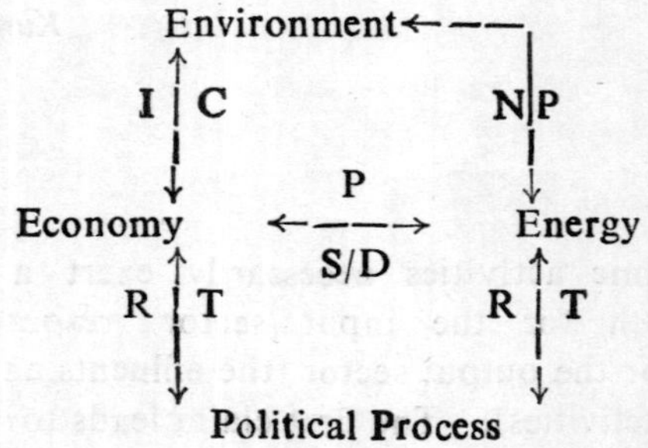

Fig. 1. ***Interdependencies between Economy, Environment, and Energy Base***

Traditional Economic Approach

The traditional Economics is the study of how society can allocate scarce resources to satisfy human needs. It is an adhoc system of short-term-management to satisfy human wants as quickly as possible. The type of economics we use talk and teach, has served society well in providing it with tools for short term decision-making to solve short term problems of what to produce, how to produce and for whom to produce. It has worked well over the past couple of 100 years with low level of population, apparently unlimited space, energy and other resources. Nature was able to absorb local shocks and redress or repair the global imbalance of natural forces. Men were too few and their interventions were too mild to endanger the life support system.

Man-Nature Relationship

However, hitherto almost overlooked problem has now come to the force. The problem has matured through centuries in the current decade. The man-nature relationship has changed with starting speed and effect. The scale of present population and the scale of its intervention and its impact on natural systems have grown at such a geometric rate that the historic man-nature relationship has been turned on its head[3]. In stead of balances there are imbalances and malfunctions. The problem have arisen because we are all living in a world with limited natural resources including air and water. In the attempt to produce more goods (higher GNP) we have tended to crowd in on ourselves and limit the range of activities available to our neighbours and for the posterity.

Collective Choice

In the narrow meaning the traditional economics, ignores the environment, ecology and use of scarce resources. With all planned economic events there is a series of unitended and uneconomic eventualities. With higher production and consumption of goods, there is environmental degradation, ecological imbalances and resources depletion. Therefore, the traditional approach to all economic problems must be broadened to include those activities that affect the environment, ecosystem and energy position. If economics is to be made relevant for next 500 years if not 5000 years, the economic apparatus need to be extended and modified. Because the society is the manager of a timeless system of finite and limited resources, it should not be guided by the logic or short-term economic considerations. We need a new set of principles to evaluate the long term consequences of each action undertaken at present. In short an intergenerational set of criteria for economic decision making which, concerns collective choice open to the community is urgently needed. The problems such as environmental degradation, resources depletion, etc., are shared by all members of society. Therefore, they can only be resolved within a framework of public decision. Another reason why such decision must be made within a public framework is that the individual evaluations of a collective choice may differ significantly. People may have different opinions about the value of nuisance of a particular collective choice[3], for instance health care schemes or the level of pollution. The problem, therefore, becomes one of balancing public purpose with private preferences.

Material Balance Model

In actuality we find society/consumers are interested not so much on actual goods but rather the services/utilities that the bundle of goods renders. I do not consume a fan, a car, or a refrigerator. The phrase "final consumption", which we use very frequently is misleading[4]. Because material goods do not disappear into a bottomless pit but are returned to the nature after the consumer has derived the utility from it. Physically nothing really disappear, the total mass of all produced goods at every level of production is always equal to the weight of the goods in the earlier state of production as well as to the weight of final product plus the by-products.

The production sector absorbs material inputs from the environment and transformers these inputs into final products and residuals. The final consumption generates wastes as well as residuals for recycling. Hence the law of conservation of matter and energy states that the physical masses of all inputs is identical in amount to the residual wastes. Depending on how tightly controlled the production and consumption process are, more or less all the by products considered as waste material are returned to their original environment and usually affect its quality detrimentally.[5]

Economics of Choice

The price system does not reflect the opportunity cost of

Volume of nputs = Volume of Waste + Volume of Consumption

$$I_a + I_r \equiv W_p + G_c$$

$$G_t \equiv W_c + W_r$$

$$\because \quad W_r \equiv W_{rw} + I_r$$

$$\therefore \quad G_c \equiv W_c + W_{rw} \times I_r$$

Hence

$$I_a + I_r \equiv W_p + W_c + W_{rw} + I_r$$

$$\therefore \quad I_a \equiv W_p + W_c = W_{rw}$$

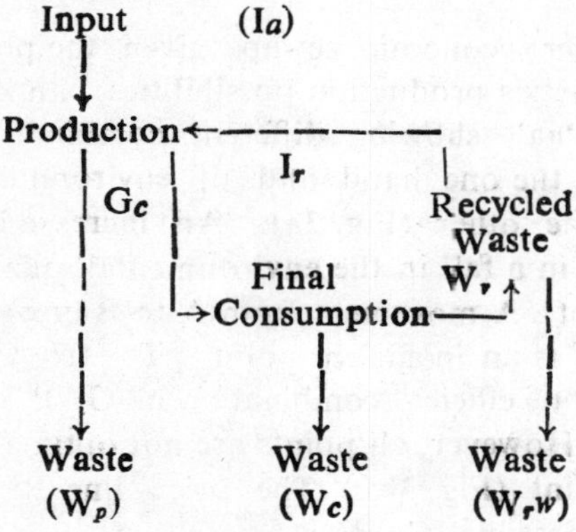

Input (I_a)

Production

G_c

I_r

Recycled Waste

W_r

Final Consumption

Waste (W_p)

Waste (W_c)

Waste (W_{rw})

Fig. 2. Material Balance Model

natural resources to the present users or to future users. It fails to convey the exact information about the relative scarcity of environmental resources to the society/users. Therefore, the economic system must encompass all the factors of production and products of a given economic activity. The welfare of an individual (or society) depends upon those goods and services, he consumes as well as many intangibles such as environmental quality. The traditional measure of welfare considers only those goods and services that are bought and sold. But economists like Nordhaus and Tobin have tried to revise the conventional measure of G.N.P. to allow for pollution cost, disamenities, leisure, recreation, shortage of natural resources etc., and they have developed a concept i.e. NEW (Net Economic Welfare) to gauge the quality of life. Environmental services have significant contribution to the welfare index. Therefore, here an economic problem arises with environmental resources. It involves a choice. The choice lies between an increase in the quantity of goods produced and a corresponding fall in the environmental quality. There is no basic conflict between economic welfare and environmental care, but the basic economic problem one encounters is how the consumer can find an optimal balance between the two.

Suboptimality in Case of Environmental Degradation

In broader economic set-up, given the present level of technology society's production possibilities can be described by the curve 'aa', showing different combinations of goods and services on the one hand and of environmental services (quality) on the other (Fig. 3a). An increase in GNP from q_1 to q_2 results in a fall in the environmental quality (unhealthy environment). A movement from A to B however improves both because A is an inefficient point. The frontier represents the possible set of efficient combination of GNP and environmental care. However, all points are not optimal. Point is a suboptimal point (Fig. 3b). The price line PL is tangential to the highest possible IC_2 at m, where consumption efficiency is ensured i.e.

$$MRS = P_e / P_g$$

There is over production of goods but under production of environmental quality. But at that point neither the price line nor the IC is tangential to the PPC aa. It means that at point m. what the consumer is willing to give up in terms of one good in order to acquire extra unit of the other is not

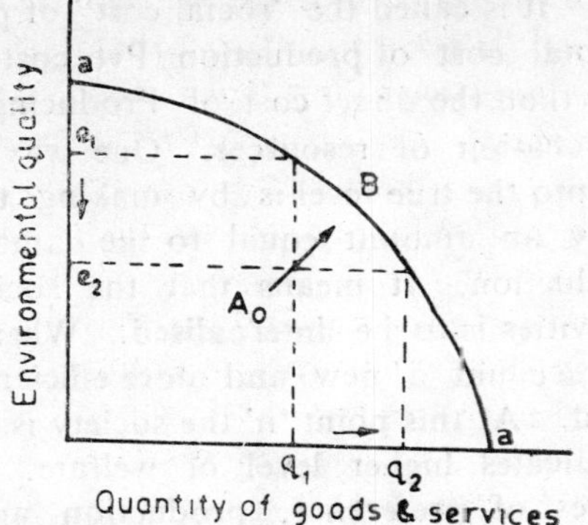

Fig. 3a. ***Set of Social Choice***

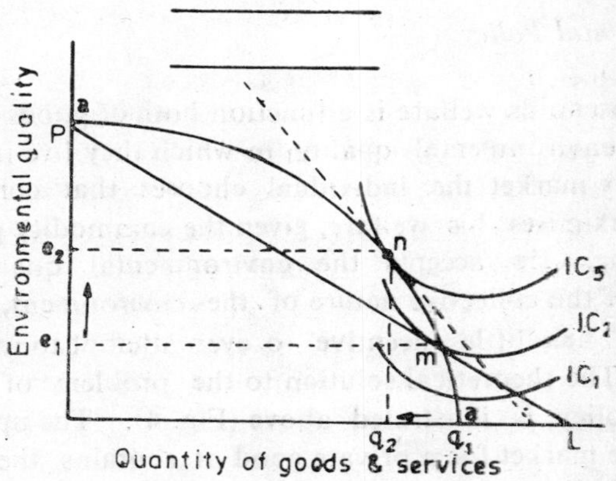

Fig. 3b. ***Resolution of Suboptimality***

equal to the cost of producing that marginal unit. The implications of this divergence of the two slopes are far reaching both from the point of view of individuals short run welfare and the society's concern for efficient allocation of resources over the long period. The slope of PL is too low. T he price line only reflects the private cost of producing goods but it does not reflect the associated cost of reduced environmental quality. It is called the "social cost" of production of goods. The total cost of production (Pvt. cost+Social cost) is much greater than the direct cost of Producing goods. So there is misallocation of resources. One way of raising the cost of goods upto the true level is by making the producers pay the society an amount equal to the damages caused in course of production. It means that the social cost of all production activities is to be internalised. When social cost is taken into account a new and more efficient equilibrium point is attained. At this point 'n' the society is on a higher IC_5, which indicates higher level of welfare. At this point three boundaries of preference, production and price are tangentially equal. Here only optimal point is attained. Tastes, costs and prices coincide and efficient resource allocation is ensured.

Environmental Policy

Households welfare is a function both of goods consumed and the environmental quality in which they live in. In the commodity market the individual chooses that consumption which maximises his welfare, given the commodity prices and his income. He accepts the environmental quality-per-se. Because of the collective nature of the environment, a single individual has little incentive to even attempt to improve its quality. The theoretical solution to the problem of environmental policy is illustrated above (Fig. 4). The upper panel depicts the market for a private good it contains the demand curve and the producer's direct marginal cost (Marginal Private cost—MPC) curve. 'm' is the usual equilibrium ensured by the narrow price system, where farms MPC=Price that equates the demand and supply forces. The lower panel of the diagram depicts that Marginal willingness to pay (MWP) increases with deteriorating environmental quality. We assume that the degree of environmental loading is directly proportional to the volume of production. Let us imagine an Effective Environmental Protection Agency (EPA) that knows

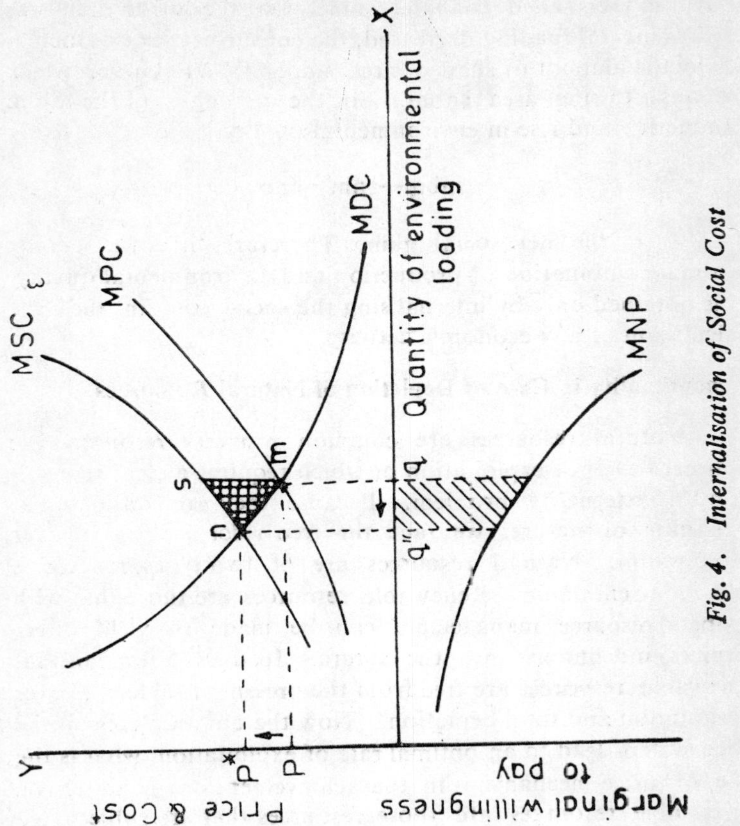

Fig. 4. Internalisation of Social Cost

the exact form of consumers MWP curve and has power to force the firms to fully compensate consumer for worsened environmental quality. Then MWP curve is to be added with MPC in order to obtain the marginal social cost (MSC). Obviously then m is no longer the equilibrium solution. Now MSC intersects the demand curve of point 'n', where the equilibrium price P* reduces the volume of supply and demand for commodity and hence the environmental loading to the point of q*. Such a fall in production leads to a decrease in combined producers and consumers surplus by the amount shown in the shaded triangle, snm. On the other hand as environmental loading decreased, the consumers surplus increases by the amount of shaded area under MWP curve, which is equal to the area snbm. So the net gain of the fall in production and rise in environmental quality is i.e.

$$snbm - snm = bnm$$

It is the net social gain. Therefore, 'n' is the socially optimal combination of production and environmental quality. It is obtained only by internalising the social cost in the cost calculation of any economic activity.

Suboptimality in Case of Depletion of Natural Resources

Natural resources are common property resources, the optimal level of exploitation of which requires a clear recognition of externality problem. Because they are considered as free gifts of nature, we face the free rider problem of over exploitation. Natural resources are of two types, renewable and non-renewable. Renewable resources are those that with proper resource management can be made to yield direct services indefinitely into the future. It does not mean that renewable resources are free from the pressing problem of over exploitation and total depletion. Now the question is : can the price system lead to an optimal rate of exploitation, what is the role of price-mechanism in the achievement of this objective. Renewable resources are those resources that are subject to a natural rate of growth and therefore, are capable of yielding a sustainable level of production into the future. The rate of their renewable directly depends on the amount of the resources left unharvested. The rate of increase of the resources consists of its rate of growth less its mortality rate i.e.

$$\triangle S_t = S_{t+1} - S_t = f(S_t) - h_t$$

It simply states that the variation in the level of stock ($\triangle S_t$) is equal to growth in stock in one period $f(S_t)$ minus the amount of harvest or hunting (ht.). Let us obtain the natural growth curve for a renewable resource. For any renewable resource (say the population of a particular animal) the rate of growth $\triangle S$ can be considered to be directly proportional to the level or stock of that population. It gives a straight line (og) (Fig. 5) function. But the expansion is in actuality halted by mortality factors, such as diseases, production and limitation in the supply of food. m curve and g curve intersect at a point, which corresponds to the maximum population (S_m) which the environment can support. When population is low, natural

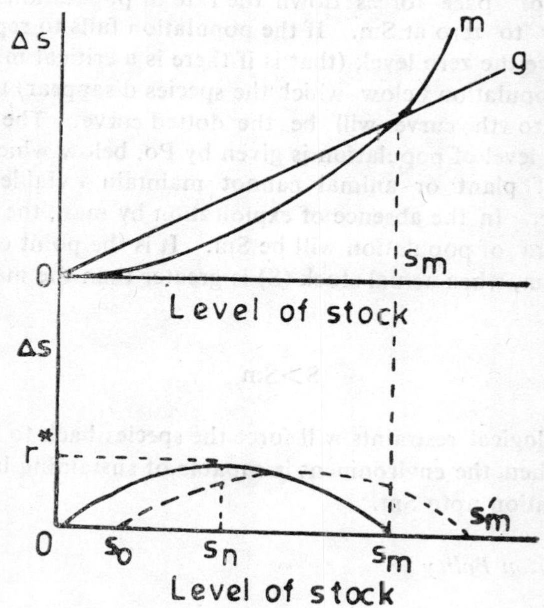

Fig. 5 Natural growth curve for a renewable resources

mortality is also low, as population increases, mortality increases at an increasing rate. By subtracting these two curves we obtain the natural growth curve of the renewable resource (OS*m*). It is concave downwards. It is of critical importance; it shows that a sustained yield can be achieved for any level of population provided the rate of harvest does not exceed the rate of change in the population i.e.

$$ht \not> f(S_t)$$

Hence sustainable yield is the maximum amount that can be harnessed from a given stock without decreasing its level in the next period the growth rate is maximum at r* level. Beyond that level, the rate of growth falls, with further rise in the population. At low level of population there is plenty of food, sufficient space etc., which results in a rapid rise in the stock. At higher levels of population, scarcity of food and relative shortage of space forces down the rate of population growth eventually to zero at Sm. If the population fails to reproduce itself above the zero level, (that is if there is a critical minimum level of population below which the species disappear) than the natural growth curve will be the dotted curve. The critical minimum level of population is given by Po, below which some species of plant or animal cannot maintain a viable rate of renewable. In the absence of exploitation by man, the natural equilibrium of population will be Sm. It is the point of stable equilibrium, when actual stock (S) is greater than the maximum stock i.e.

$$S > Sm$$

then ecological restraints will force the species back to Sm. If S<Sm, then the environment is capable of sustaining increases of population upto Sm.

Conservation Policy

Harvesting or hunting a natural resources requires money and effort. So it is associated with positive cost. So one can state that the level of stock depends on the extent of harvest which itself depends on the level of effort and that level of effort determines the level of cost. Higher effort represents higher cost. Let us derive the sustainable yield curve (SYC) from the natural growth curve and super-impose it on the cost curve Cc. (Fig. 6). Both intersect at K point, H_1 in the amount of Harvesting effort. Here AC of harvest is equal to the average yield.

i.e. AC=AY

Here marginal cost is not equal to the marginal yield. So 'K' is not the efficient point. Each individual operate independently. As long as profit exists, harvesting will continue. The resources are over exploited. In order to correct this misuse (over use) of the resources, several alternative methods should be adopted by the public authority.

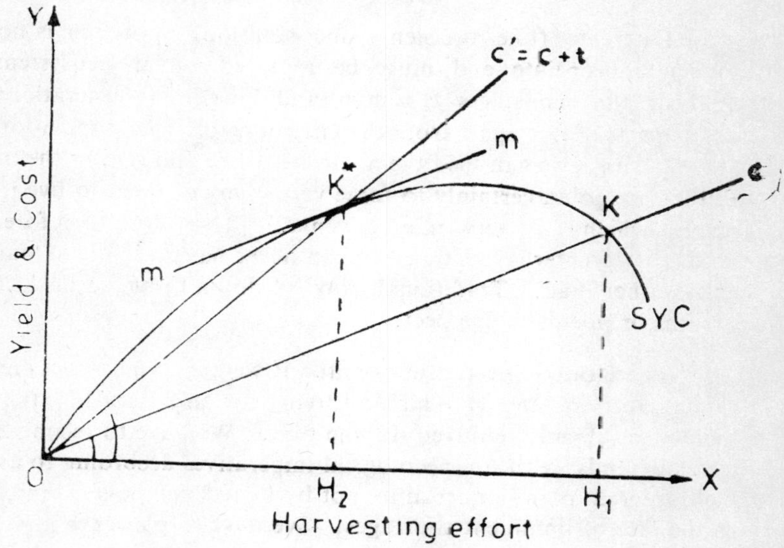

Fig. 6. Optimal point of Exploitation of renewable resources

It can increase the cost of harvesting by imposing a tax on the level of harvesting effort. The cost curve then shifts to a higher level. It should be raised to the level of efficient equilibrium point i.e. K*. At this point slope of the sustainable yield curve is equal to the slope of total cost curve i.e.

MC=MY

This efficiency condition necessitates the harvesting effort to fall to the level of H_2. Without any public regulation, the optimal level of effort/harvest is not achieved. There is over exploitation. Therefore, the intervention of the public authority is warranted for internalising the externality of common properly resource. Apart from licensing and tax, regulation should include the size of hunting/harvest, limitation on seasons and equipments used in harvesting process etc.

Niche Specification

Discreate (one problem—one solution) approach is no longer appropriate and must be replaced by an ecosystem analysis that considers ths man and his coming generations as a part of not apart from the environment. As population keep soaring, we can feed them, clothe them, and shelter them. But we are going certainly to force very many of them to live in niches which are very non-congenial.[7] Thc future therefore, holds ever greater restrictions we can never have a living as our grandfather had. Traditional way of doing thing, seems like dreams or poems of the pect.

Therefore efforts are essential if we are to maintain any resemblance of the standard of living (or quality of life) that has been already enjoyed in the past. We have to reconcile our demands with the ecological imperatives according to an well ordered plan but certainly not by panicky impoverishment in the face of imminent shortages. We have to place the maintenance and preservation of the fragile ecosystem ahead of technology, population and GNP.

Broad Economic Horizon

Fresh breathable air, dr nkable water, pleasant environment, sustainable energy base, appears to become relatively scarce commodity. Since economics is particularly concerned with the items of scarcity it is obvious that environmental quality and resource depletion can benefit by the use of more scientific economic theories and methods. Therefore, the traditional approoch to economic problems must be broadened to include all those activities that affect the environment and energy base.

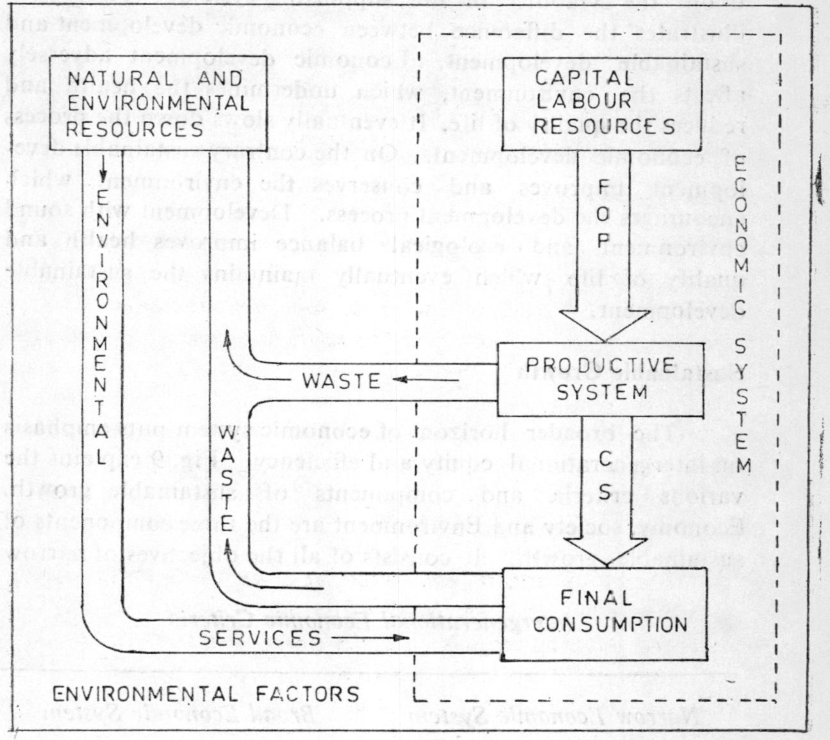

Fig. 7. Environment Vs. Narrow Economic System

An analysis of the capability of the economic system for resolving economic problems of allocating scorce resources to differing and competing ends must include both the flow of natural and environmental resources into the production process and the flow of wastes from the production and consumption activities back to the natural environment. Equally important are the services that consumer directly obtain from the natural environment. The Fig. 7 illustrates the economic system with a broader horizon. The environmental and natural resources directly affects the standard of living or quality of life of people. The welfare of the individual is ultimately dependent upon the viability of life supporting system.[6] The Fig. 8 illustrates the difference between economic development and sustainable development. Economic development adversely affects the environment, which undermines the health and reduces the quality of life. It eventually slows down the process of economic development. On the contrary sustainable development improves and conserves the environment, which encourages the development process. Development with sound environment and ecological balance improves health and qualify of life which eventually maintains the sustainable development.

Sustainable Growth

The broader horizon of economic system puts emphasis on intergenerational equity and efficiency. Fig. 9 explains the various criteria and components of sustainable growth. Economy, society and Environment are the three components of sustainable growth. It consists of all the objectives of narrow

Intergenerational Economic Criteria

Narrow Economic System	*Broad Economic System*
Present	Future
Production	Protection
Quantity	Quality
Vertical Growth	Sustainable Growth
Consumption	Conservation

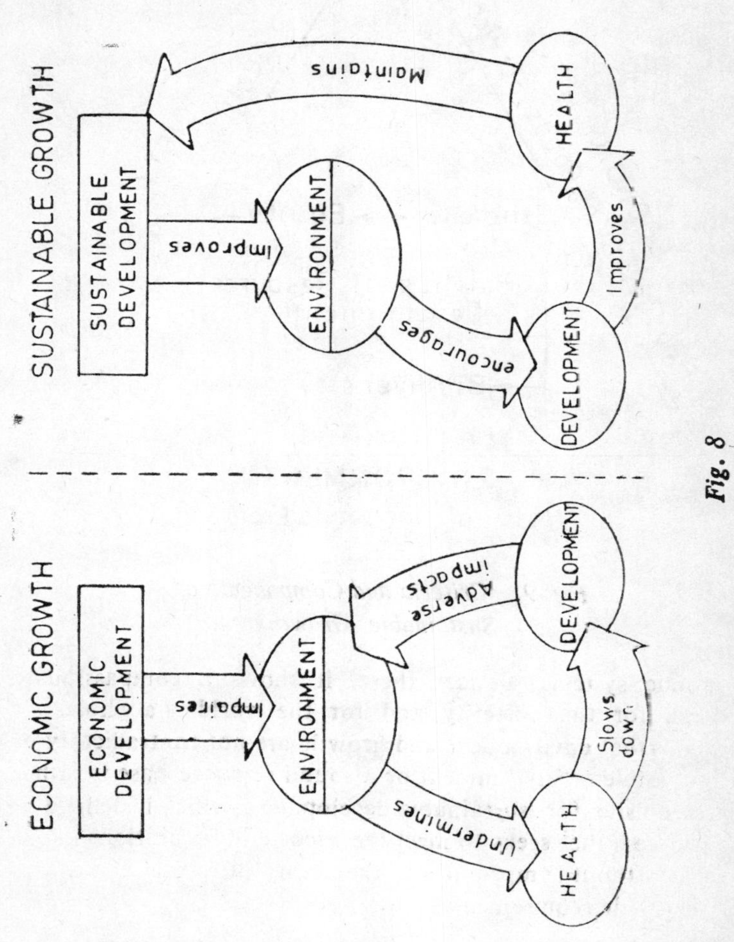

ECONOMIC GROWTH
ECONOMIC DEVELOPMENT
Impaires
ENVIRONMENT
Undermines
HEALTH
Slows down
DEVELOPMENT
Adverse impacts
SUSTAINABLE GROWTH
SUSTAINABLE DEVELOPMENT
Improves
ENVIRONMENT
encourages
DEVELOPMENT
Improves
HEALTH
Maintains

Fig. 8

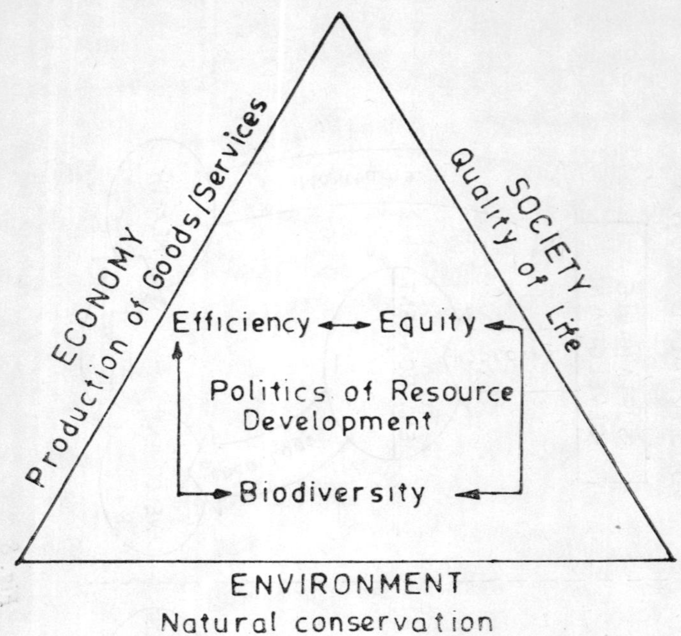

Fig. 9. Criteria and Components of Sustainable Growth

economic system. Besides these, it shows a compassionate concern for the posterity and for the world as a whole. It contends that environment and growth are not contradictory or incompatible. Environment or a sound resource base is rather a prerequisite for sustainable development which is defined as the progress that seeks to meet the needs and aspirations of the present without compromising the ability of future generations to meet their requirements.

Environment protection and resource conservation and economic development are to sides of the same coin. Development can not subsist upon a deteriorating resource base. Development process should be less "energy-intensive" and more sensitive to environmental impacts. No country, no economic agent should be allowed to exploit its comparative advantage in 'pollution-intensive' or 'energy-intensive' goods. So the broader system should deal with environmental, energy and developmental issues together. Our common future now has served a notice that the time has come for a marriage between economy and ecology, so that the Government and people can take responsibility not just for environmental damages but for the policies that cause such damages.

Intergenerational Criteria

Ecologically over stressed society is further impelled to cause more intensification of pressure on its environment in an attempt to survive. Now at least we have a rude awakening and a jolting realisation of what have been done to the natural environment and how many tends to jeopardise its future. The society at present and in future. This is the only approach which can ensure a sound and sustainable balance between a productive environment and a protective environment. In view of protection, and preservation, the negative effect of the rise in material welfare requires reorientation with respect to certain prevailing economic tendencies so that in addition to economic development, the protection of environment and preservation of natural resources have to be conceived of as an explicit aim in planning and in political decision-making. An integrated planning framework has been illustrated in Fig. 10. Conservation objectives are as important as Development objectives in order to determine the policies and strategies for sectoral and spatial planning. Sectoral and spatial planning along with environmental impact Assessment influences the implementation of development activities which should be followed by auditing and evaluation. After auditing and evaluation, through project redesign or process reforms the conservation and development objectives and to be redefined. Hence both the 'need to produce' and 'need to peotect' are vitally relevant and indispensable, neither should be compromised. Since no policy can pretend to affect one without influencing the other, both the needs must be taken into account in the policy

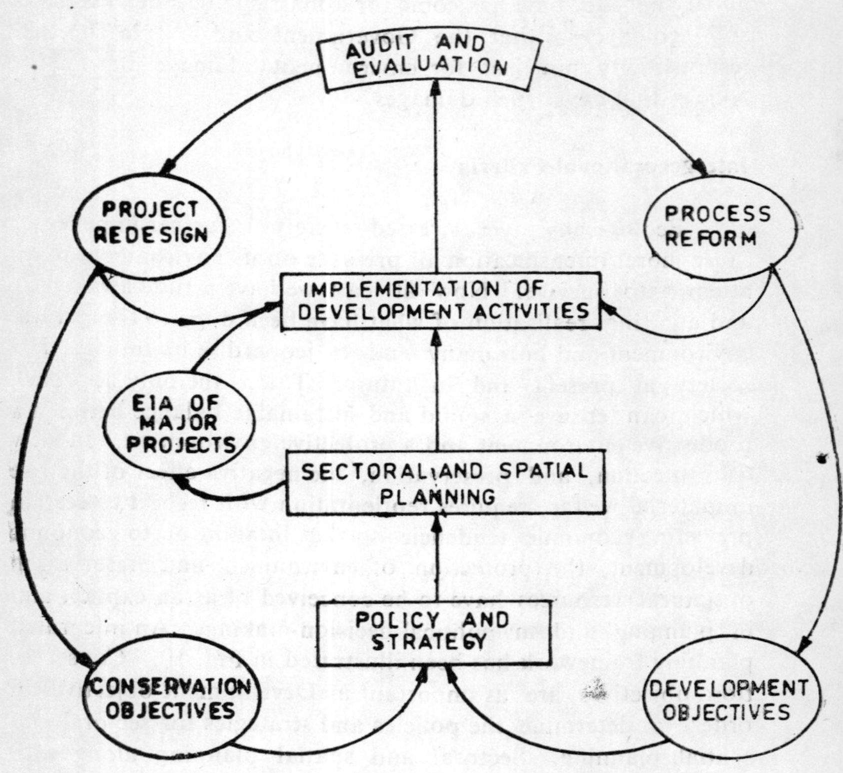

Fig. 10. Integrated Planning for Growth Sustainability

formulation process and balanced in the ultimate decision making. Notwithstanding the practical difficulties we should be concerned about the problem in its proper dimension and severity. If we really wish to improve the quality of life for us and for the posterity, and not just talk about doing so, we must apopt appropriate public policies that sacrifice at present some ordinary growth of GNP to enjoy the bencfit of higher NEW (Net Economic Welfare).

Notes

1. Smith, (1976), p. 5.
2. Das, (1985), p. 284.
3. Waugh, G., (1979), p. 325.
4. Hjalte, (1977), p. 2.
5. Nijkamp, (1977), p. 11.
6. Das, (1988), p. 50.
7. Murphy, (1967), p. 3.
8. *Ibid.*, p. 28.

References

Breese Gerald, (1978), *Urbanisation in newly Developing Countries.*

Das Kumar B, (1980), The cost of Urban & City Life in developing countries, Presented at the Advanced *Summer Institute held at Vrije University, Amsterdam,* (Holland), during August.

———, (1986), Urban cost of spatial concentration in developing countries (inequity and inefficiency aspects), Presented at the European Summer Institute, *held at University of Umea, Sweden,* June.

———, (1985), Distortion of Social Welfare Function, *The Indian Journal of Economics,* Vol. LXVI, No. 261.

———, (1988), Ecosystem approach for sustainable economic Development, *Indian Journal of Regional Science,* Vol. XX (i).

Das Kumar B., (1990), Intergenerational set of Criteria for sustainable Growth in developing countries, paper presented in the Advanced Summer Institute on Environmental Management, held of University of Aberdeen; U.K. during August.

Dasmann R., (1968), *A different kind of country,* Macmillan, New York.

Dubos Rene, (1968), *Human and Animal,* Scribner, New York.

Gadgil M., (1985), Towards an Ecological History of India, *Economic & Political Weekly*, Vol. XX, No. 45-47, pp. 1909-18.

Murphy, E. P., (1967), *Governing Nature*, Quardrangle Books, Chicago.

Mishan E. J., (1967), *The Cost of Economic Growth*, Staples Press, London.

Nijkamp P., (1977), *Theory and Application of Environmental Economics*, North Holland.

Odum, E. P., (1976), The strategy of Ecosystem development in the *Ecology of Man (ed.)*, R. L. Smith, Section-III, Harber & Row.

Ridkar R. K., (1973), To grow and not to grow thats not the relevant question, *Science*, 1982, pp. 1315-18.

Turvey R., (1963), On divergence between social Cost and Private Cost, *Economics*, August,

Werner J., (1974), Environmental Protection and Problems of Growth, *Economics*, Vol. 9.